Electric Drive System Design for Electric Vehicles

Electric Drive System Design for Electric Vehicles bridges the gap between specialized research on electric vehicle (EV) electric drive system design principles and developing industry-driven solutions. It addresses best practices for ensuring the performance and reliability of EV electric drive systems, including battery, motor, and power electronic components.

Based on the author's hands-on experience, the book takes a multidisciplinary approach to EV drive system design, combining electrical engineering, thermal design, mechanical engineering, and manufacturing expertise to deliver efficient, reliable, and high-performance solutions. It includes case studies and practical examples from the industry, reviewing state-of-the-art electric system technologies in current EVs such as Lucid, Tesla, Chevrolet, VW, Nissan, and more. Readers will also obtain insight into how fundamental physics plays a critical role in those technological advancements and innovations.

This book will benefit academic researchers and graduate students studying power electronic packaging and electric drive systems. It will also serve industry professionals involved in EV design and general power electronic system packaging.

Yunqi Zheng has 20 years' experience with system integration, design, and thermal/mechanical/electromagnetic design optimization on electric drive systems. Currently, Dr Zheng is a system engineer working on electric drive systems in Honeywell Aerospace Technologies, USA. She holds a PhD/MSc in Mechanical Engineering from the University of Maryland College Park, USA, and an MSc/BS in Engineering Mechanics from Tsinghua University, China. Dr Zheng was a core technical member contributing to the success of in-house power electronics design for an EV start-up. She was also one of the lead engineers for an ultra-small EV on-board charger system prototype. Previously, Dr Zheng was with GM Electrification, working on design/analysis and system integration for next-gen power inverter systems. She holds several patents and has written several publications.

Electric Drive System Design for Electric Vehicles

Yunqi Zheng

CRC Press
Taylor & Francis Group
Boca Raton London New York

CRC Press is an imprint of the
Taylor & Francis Group, an **informa** business

Designed cover image: Shutterstock

First edition published 2026
by CRC Press
2385 NW Executive Center Drive, Suite 320, Boca Raton, FL 33431

and by CRC Press
4 Park Square, Milton Park, Abingdon, Oxon, OX14 4RN

CRC Press is an imprint of Taylor & Francis Group, LLC.

ISBN: 978-1-032-94290-2 (hbk)
ISBN: 978-1-032-94291-9 (pbk)
ISBN: 978-1-003-56996-1 (ebk)

DOI: 10.1201/9781003569961

Typeset in Times
by Deanta Global Publishing Services, Chennai, India

To my family, Zaifeng, Mingfei, and Mingteng.

Special thanks to our parents, 家彬 *and* 昆川, 行琦 *and* 淑玲, *who have supported us all these years.*

Contents

About the Author xii
Preface xiii
List of Abbreviations xv
Acknowledgments xvii

Chapter 1 Introduction to EV Electric Drive Systems 1

Introduction 1
Battery 1
Motors 2
Inverter 3
Transmission 3
OBC 4
DC-DC Converter 4
Thermal Management System 4
Bibliography 7

Chapter 2 EV Performance Characteristics and Electric Drive Systems 8

Introduction 8
Driving Range 8
Torque 10
Power 13
Acceleration 15
Top Speed 16
Charging 17
Efficiency 17
Bibliography 19

Chapter 3 EV Traction Motor and Thermal Design 22

Introduction 22
Permanent Magnetic Synchronous Motor 22
  Stator winding 24
  Rotor Magnetics 25
  Laminations 26
Induction Motor 26
Motor Parameters 27
  Torque and Power 27
  Speed 30
  Efficiency 31
Thermal Management 34

Motor Loss 34
Thermal Resistance 36
Liquid Cooling 37
Coolant 38
Housing Jacket Cooling 39
Spray Cooling 39
Shaft Cooling 40
Other Cooling Methods 40
Summary 41
Bibliography 42

Chapter 4 Battery System Packaging and Thermal Management 45

Abstract 45
Basics of Lithium-Ion Batteries 45
Charging and Discharging 46
Battery Loss and Resistance 47
Battery Cell 47
Battery Performance Characteristics 48
Battery Management System 50
Battery System Packaging 51
Thermal Runaway 52
Battery System Thermal Management 54
Thermal Management Objective 54
Heating 55
Bottom Cooling 55
Side Cooling 56
Comparison between Two Battery System Cooling Methods 57
Technology Trend 57
Bibliography 58

Chapter 5 Thermal Design of Power Electronic Components 61

Abstract 61
Introduction 61
Power Electronic Components 61
Inverter 61
OBC 64
DC-DC Converter 65
Efficiency and Power Loss 66
Power Electronic Component Thermal Design 68
Thermal Design for Power Semiconductor Devices 68
TO220- and TO247-Based Discrete Power Electronic Packages 69
DPAK/D2PAK Discrete Power Electronic Packages 71

Multichip Power Module ... 74
Case Study ... 76
Transient Performance ... 78
Trend ... 79
Discrete Power Electronic Packages ... 79
Double Side Cooling Power Module ... 79
Bibliography ... 81

Chapter 6 Power Electronic Packaging Interconnects and Thermo-Mechanical Reliability ... 85

Introduction ... 85
Wire Bonding ... 86
Die Attach and Heat Sink Attach ... 88
Stage 1. Preheat ... 89
Stage 2. Soak ... 89
Stage 3. Reflow ... 89
Stage 4. Cooling ... 90
Thermo-Mechanical Fatigue ... 91
Power Cycling Test and Temperature Cycling Test ... 93
Thermo-Mechanical Fatigue Model ... 96
Trend ... 98
New On-Chip Interconnects ... 98
Ag Sintering ... 98
Pressfit ... 100
Transfer Molded Package ... 101
Bibliography ... 101

Chapter 7 EV Power Electronic System Vibration ... 106

Introduction ... 106
Factors Affecting Vibration Reliability ... 106
Excitations ... 106
Resonance ... 108
Components Vibration Resistance ... 108
Vibration Characterization ... 109
PSD ... 109
ISO Standards ... 109
ISO 19453-3 ... 110
ISO 16750-3 ... 111
Mission Profile ... 111
How to Design Vibration Proof Components ... 113
Increase the fundamental frequency of the PCB ... 113
Frequency Separation ... 114
High vibration-resistant components ... 116

Damping ... 116
Other methods ... 117
Methodology for Vibration Design Optimization/Validation ... 117
Bibliography ... 117

Chapter 8 EMI in Electric Drive Systems ... 119

Abstract ... 119
Introduction ... 119
EMI Types ... 120
Conducted EMI ... 120
Common Mode EMI ... 121
Differential Mode EMI ... 121
Radiated EMI ... 121
Electromagnetic Compliance (EMC) ... 123
Factors Affecting EMI ... 124
Devices and Gate Drive Circuits ... 124
Parasitic Components ... 125
Critical Parasitic Loop Inductances in Converters ... 125
Critical Parasitic Loop Inductance in Inverter ... 126
EMI in Power Converters ... 127
EMI in Inverters and Motors ... 128
Resonance ... 130
Cable ... 130
Cable Radiation (Antenna) ... 131
Cable Resonance ... 131
Other EMI Sources in EVs ... 131
EMI Mitigation ... 132
Grounding ... 132
EMI Filters ... 132
PCB and System Layout ... 134
EMI Mitigation in Cable Harness ... 134
Resonance ... 134
EMI Measurement Methods and Tools ... 135
Bibliography ... 136

Chapter 9 Trends in EV Electric Drive Systems ... 139

Abstract ... 139
Integration and Modularization ... 139
SiC ... 140
Why SiC MOSFET? ... 141
State of Art of SiC Packages ... 143
800 V HV Electrical Architecture ... 144
Introduction ... 144

800 V SiC inverter 146
800 V Battery 146
800 V Motors 147
800 V OBC 147
800 V Capacitors 147
Cables 148
Bibliography 148

Index 151

About the Author

Yunqi Zheng has 20 years' experience with system integration, design, and thermal/mechanical/electromagnetic design optimization on electric drive systems. Currently, Dr Zheng is a system engineer working on electric drive systems in Honeywell Aerospace Technologies, USA. She holds a PhD/MSc in Mechanical Engineering from the University of Maryland College Park and an MSc/BS in Engineering Mechanics from Tsinghua University. Dr Zheng was a core technical member contributing to the success of in-house power electronics design for an electric vehicle (EV) start-up. She was also one of the lead engineers for an ultra-small EV on-board charger (OBC) system prototype. Previously, Dr Zheng was with GM Electrification working on design/analysis and system integration for next-gen power inverter systems. She holds several patents and has written several publications.

Preface

Adoption of Electric Vehicles (EVs) is one major technology trend that emerged in the last 10 years. 'Electric cars accounted for around 18% of all cars sold in 2023, up from 14% in 2022 and only 2% 5 years earlier, in 2018' (IEA, 2024). (Goldman Sachs, 2024) predicted that EVs could make up as much as nearly half of global car sales by 2035. The electric drive system is the heart of EVs. It is the key system that makes EVs different from traditional gasoline vehicles with internal combustion engines (ICEs).

Electric drive system design is the system design and integration of battery pack, traction motors, inverter, and other high voltage (HV) components such as on-board chargers (OBCs) and DC-DC converters. EV drive system design requires a multidisciplinary approach combining power electronics, mechanical engineering, thermal design, and manufacturing expertise to deliver efficient, reliable, and high-performance solutions. There have been numerous research publications about various advancements and innovations in the area such as double-side cooling, more integrated systems, use of advanced joints, etc. Yet no book has systematically explained those advancements and innovations. The books available are mostly about electrical design. Even a few covered electric drive systems, but outdated examples were used. It is rarely explored how fundamental physics drives those state of art developments.

The aim of the book is to bridge the gap between specialized research on EV electric drive system design principles and industry best practise. Also, the book intends to address the topic from a more balanced multidisciplinary point of view in addition to electric engineering. The intention is to showcase many engineers' excellent hands-on work and to provide readers with information on how to design an electric drive system based on a veteran engineer's field experiences.

This book is a real engineering design book. It is based on a wealth of open-sourced design data from real-world EVs including Tesla Roadster. The book will review state of art electric drive system packaging technologies in many current EVs, including Lucid, Tesla, Chevrolet, VW, Nissan, and more, some of which the author has either analyzed or worked on. Readers will obtain insight on the critical role of physics in those technological advances and innovations.

The book is suited for readers from a range of engineering disciplines including electrical engineering, automotive engineering, and mechanical engineering. The book will benefit students preparing for a career in electric mobility, or professionals seeking to advance their expertise and to innovate cutting-edge solutions in this area, or entrepreneurs looking to understand the fundamentals and to facilitate and capitalize on the electric mobility trend.

Power capability dictates EV performance and efficiency. Power capability of an electric drive system is ultimately limited by efficiency and thermal performance. Designing a system with optimum efficiency is achieved through a multidisciplinary approach. The options to increase thermal performance include reducing thermal resistance, increasing heat spreading, use of heat sink, and double-side cooling.

Those topics will be explained in detail, with examples from several leading EVs in the following chapters.

It took 100 years to mature ICE technology. How can engineers develop a reliable electric drive system in a short period of time? What is the critical technology advancement? In addition, how can engineers ensure the systems are able to survive 10 years/100,000 miles of life or even longer? EV components such as inverters and OBCs experience very fast and high switching under high power, which could potentially produce a significant amount of broadband electromagnetic interference (EMI). How to ensure this won't be an issue? This book will introduce the industry's best practice to address those concerns in several critical areas:

1. Efficiency and thermal performance
2. Design for thermo-mechanical durability
3. Design for vibration durability
4. Electromagnetic Interference
5. Critical technologies and trend

REFERENCES

Goldman Sachs. (2024, FEB 16). *The future of four wheels is all electric*. Retrieved from https://www.goldmansachs.com/intelligence/pages/the-future-of-four-wheels-is-all-electric.htm

lIEA. (2024). *Global EV Outlook 2024*. Retrieved from https://www.iea.org/reports/global-ev-outlook-2024

List of Abbreviations

EV: electric vehicle
HEV: hybrid electric vehicle
DC: direct current
AC: alternate current
OBC: on-board charger
PFC: power factor correction
BMS: battery management system
HV: high voltage
LV: low voltage
PMSM: permanent magnetic synchronous motors
PWM: pulse width modulation
EMF: electromotive force
ICE: internal combustion engine
WOT: wide open throttle
SPM: surface permanent magnet
IPM: interior permanent magnet
RPM: revolutions per minute
HTC: heat transfer coefficient
WEG: water/ethylene glycol
ATF: automatic transmission fluid
DCIR: direct current internal resistance
ACIR: alternating current internal resistance
Ah: ampere-hours
SoC: state of charge
SEI: solid electrolyte interface
EMI: electromagnetic interference
DAB: dual active bridge
WBG: wide band gap
CM: common mode
DM: differential mode
PCB: printed circuit board
IMS: insulated metal substrate
FR4: flame retardant 4
DBC: direct bond copper
DBA: direct bond aluminum
TIM: thermal interface material
CTE: coefficients of thermal expansion
PSD: power spectral density
OEM: original equipment manufacturer

MLCC: multilayer ceramic capacitors
EMC: electromagnetic compliance
CISPR: International Special Committee on Radio Interference
HF: high frequency
LISNs: line impedance stabilization network

Acknowledgments

I am deeply grateful to my publisher, Taylor & Francis Group/CRC Press for their belief in the project. I express my sincere appreciation to the book editors, Kyra Lindholm and Sonia Tam, for helping me through the journey.

1 Introduction to EV Electric Drive Systems

INTRODUCTION

In electric drive systems (see Figure 1.1), the battery provides the power to drive the vehicle. The power from the battery is high voltage (HV) direct current (DC), which is converted to alternate current (AC) power by the inverter. The AC power then is outputted to the motor. The main function of the motor is to convert AC power to mechanical power, which is passed through the transmission to drive the wheels. During braking and deceleration, the motor can also serve as a generator, recovering the mechanical power from the wheels into electrical power to charge the battery.

Inverters manage power from the battery, commanding the motor to deliver torque and speed precisely by regulating the frequency and voltage of the AC power.

On-board chargers (OBCs) charge the battery using the AC power from the grid. The AC power is converted to the DC power suitable for the battery system. DC fast charger is another charging option, which provides high DC power around 50–350 kW directly to the battery.

Another key power electronics component is the DC-DC converter. Its main function is to convert high DC voltage (400–800 V) from the battery into low DC voltage (48 or 12 V) to power low voltage (LV) electrical components like control circuits, lights, wipers, and pumps in EVs. LV power is needed to drive all the control circuits in the EV. Those control circuits are the drivers to control all the HV components including the inverter, BMS, and OBC.

BATTERY

EV batteries are the 'gas' for EVs. To drive the vehicle, the traction motor converts electrical energy stored in the battery to mechanical energy. The capacity of the battery system directly determines the range of the vehicle, and the power from the battery determines how much power the electric drive system can deliver.

Battery capacity, a key parameter for the battery pack, is typically expressed in kilowatt-hours (kWh), representing the battery's maximum energy storage capacity. An EV's battery capacity is analogous to the size of a full tank of gas in an internal combustion engine (ICE) vehicle. EV battery capacities typically range from 40 to 100 kWh; the higher the capacity, the more miles it may drive in one charge.

Lithium-ion batteries are the most commonly used batteries for EV traction applications. This is because lithium-ion batteries have relatively high energy density, high power density, and a longer cycle life than other battery types (here, cycle

DOI: 10.1201/9781003569961-1

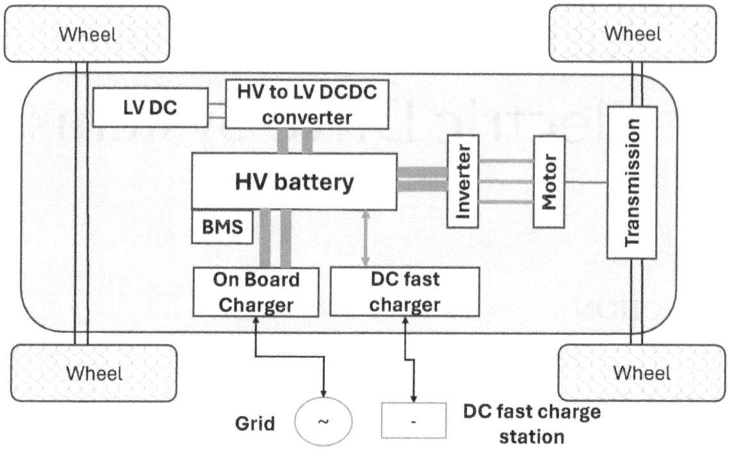

FIGURE 1.1 Overview of HV electric drive system in EV (BMS is battery management system).

life means how often the battery can be recharged to its full capacity) (US DoE Alternative Fuels Data Center, n.d.; Rajashekara, 2013).

MOTORS

Designed for EV propulsion, traction motors convert electric power to mechanical power with the commanded torque and speed to drive the vehicle. The torque and speed of the motor directly decide the torque and speed a vehicle can deliver. The motor is the key component related to several critical EV performance characteristics: torque, acceleration, speed, and power. Most traction motors are either permanent magnetic synchronous motors (PMSM) or AC induction motors. Both need AC supplied by inverters. The goal of traction motor design is to provide excellent performance without sacrificing efficiency.

One key performance characteristic is motor torque. Motor torque quantifies the rotational force that a motor can produce. It directly determines the vehicle wheel's angular acceleration and towing capacity. Mathematically, angular acceleration (a) is determined by torque (T) and moment of inertia (I) of the vehicle:

$$a = T / I \tag{1.1}$$

Torque generated by the electric motor determines how quickly the vehicle can overcome inertia and accelerate. Higher torque allows faster acceleration, resulting in a more responsive driving experience. Higher torque is normally related to higher power.

$$Power = Torque \times Angular\ Speed \tag{1.2}$$

Towing capacity refers to the maximum weight a vehicle can safely tow. Torque produced by the electric motor contributes to the rotational force needed to move the vehicle and its attached load. More torque is beneficial for towing heavier loads, as it provides the necessary force to overcome the resistance and inertia associated with towing.

The speed of an EV is decided by its traction motor speed, tire diameter, and gear ratio if a transmission with gears is used.

$$Vehicle\,Speed = Motor\,Speed \times Gear\,Ratio \times Tire\,Diameter \tag{1.3}$$

In PMSM motors, the motor speed mainly depends on AC current frequency regulated by the inverter and the motor's pole number (see Equation 1.4).

$$Motor\,Speed = \frac{120 \times AC\,Frequency}{Pole\,Number} \tag{1.4}$$

Similarly, induction motors' rotor speed is also mostly determined by AC frequency and pole numbers, except that the speed will be slightly lower due to slippage. The speed difference between the induction rotor and the magnetic field is typically around 5%. Motor speed also depends on the yield strength of the rotor structure, the bearings' life, as well as the iron and winding losses.

INVERTER

The traction inverter converts DC power from the battery into three-phase AC power. EV motors use the AC power to set up a rotating magnetic field to drive the motors. Inverters play a critical role in controlling an EV traction motor's torque and speed by precisely adjusting voltage and frequency of the AC current.

A pulse width modulation (PWM) scheme is used in the inverter to produce a sinusoidal voltage output with target voltage and frequency. With PWM control, the inverter chops DC voltage into a series of pulses by turning on/off sets of power devices very rapidly. By varying the duration of these pulses, the inverter produces a smooth AC waveform, allowing precise and instantaneous control with high efficiency (>95%) and compact inverter design. The typical inverter switching frequency for EV/hybrid electric vehicle (HEV) applications is in the 10–25 kHz range. AC frequency of the motor and number of poles for the motor determine the motor speed (see Equation 1.4). AC frequency is typically less than 1/10 of the switching frequency (around 1 kHz) (Husain, 2021).

TRANSMISSION

Most EVs use single-speed transmission. The single-speed gear reduces the speed of the electric motor to the designated speed for the wheels. In comparison, multi-speed transmission is a default choice for gasoline vehicles, because the ICE only operates most efficiently within a narrow range of speeds. It only has less than half the efficiency if its speed is outside of its optimum speed range. Gasoline vehicles can have a significant boost (~30%) of their efficiency by using multi-speed transmission. Multi-speed transmission allows the engine to stay within this optimum speed range by adjusting the gear ratio differently for speeds beyond this range. In contrast, EV drivetrains mainly comprise the motor, inverter, and battery. The efficiency variation across the broad speed range is minimal, maybe 2% combined. Therefore, for EVs, multi-speed transmission isn't needed to gain additional efficiency, at the price of increasing transmission complexity (Sgraham, 2021; Ehsani, 2003).

OBC

OBCs convert AC power from the electric grid to DC power to charge the battery. OBCs control the current and the voltage of the output power to the battery. The power ranges from 1.3 to ~22kW. Bi-directional OBCs not only charge the battery, but also allow battery power to be returned to the grid. It is emerging as one major choice for EVs (Kampl, 2020; Dhage, 2022; Yuan, 2021). AC input for OBCs can be Level 1 or Level 2 AC. Either one- or three-phase AC from the grid goes through EV supply equipment (EVSE) via a connector to an OBC. DC output can be between 200 and 900 V depending on the battery pack design.

DC-DC CONVERTER

There are multiple DC-DC converters in an EV. The main DC-DC converter converts DC power from an on-board 200–800 V HV battery to lower voltage (48 or 12 V) power (Panasonic, 2023). In addition to HV components, there are numerous LV components in EVs. It includes steering, sensors, headlights, wipers, and all the control circuits. One example is the LV power used in the inverter digital signal processor (DSP) board and gate drivers. LV components often rely on power from LV lead batteries, which need to be charged periodically by the DC-DC converter from the HV lithium-ion battery. For EVs, HV needs LV to power all the control circuits. LV is just as critical as HV and those that are LV need the power provided by the DC-DC converter.

THERMAL MANAGEMENT SYSTEM

The electric drive system generates heat due to power loss. The goal of thermal management is to dissipate the heat generated by the electric drive system and keep the system within optimum operating temperature range.

The efficiency of EVs is the percentage of the energy used to drive the vehicle (*E_out*) compared to the energy (*E_*in) provided by the battery (see Equation 1.5a). The difference between *E_in* and *E_out* is the total power loss (see Equation 1.5b).

$$Efficiency(\%) = \frac{E\_out}{E\_in} \tag{1.5a}$$

$$Power\ Loss = E\_in - E\_out \tag{1.5b}$$

EV traction power is roughly around 80–200 kW (Husain, 2021). Electric drive systems are very efficient and losses are only a couple percent. Efficiency of the battery, motor and inverter is typically higher than 90%. But even a couple percent of the 100 kW is significant. More importantly, motors and power electronic systems need to be as compact as possible to achieve high power density. Without an appropriate cooling system to remove heat, those components will overheat and fail. Keeping components within a normal operating temperature range is essential for reliability. For example, the battery needs to be operated at a narrow temperature range of 10–35°C to ensure its performance, safety, longevity, and reliability.

Liquid cooling systems are the default choice for EV batteries, traction motors, inverters, OBCs, and DC-DC converters. Performance of cooling systems can be characterized by thermal resistance (R_{th}). It indicates how effectively the heat is dissipated. It is defined by the temperature difference ($T1-T2$) between two locations divided by the heat (power loss) flow through those two locations:

$$R_{th} = \frac{T1 - T2}{Power\ Loss} \tag{1.6}$$

T2 is normally referred to as ambient temperature or coolant temperature.
T1 is component temperature.

Low thermal resistance is desired to obtain lower temperature rise under a specific amount of heat (power loss) or to pass more power and dissipate more power loss with the same temperature rise.

There are three modes of heat dissipation: conduction, convection, and radiation. Conduction and convection are common in EV electric drive system cooling. Conduction is the direct heat transfer through a material without any material movement. Convection is the heat transfer via either fluid or gas flow.

Thermal resistance for conduction (shown in Figure 1.2) is characterized as:

$$R_{th} = \frac{L}{K \times A} \tag{1.7}$$

K: thermal conductivity
A: cross-section area for heat conduction
L: heat dissipation length

Based on Equation 1.6, low conduction resistance can be achieved by reducing L, increasing A, and choosing a material with higher K.

Thermal resistance for convection (shown in Figure 1.2) is:

$$R_{th} = \frac{1}{h \times A} \tag{1.8}$$

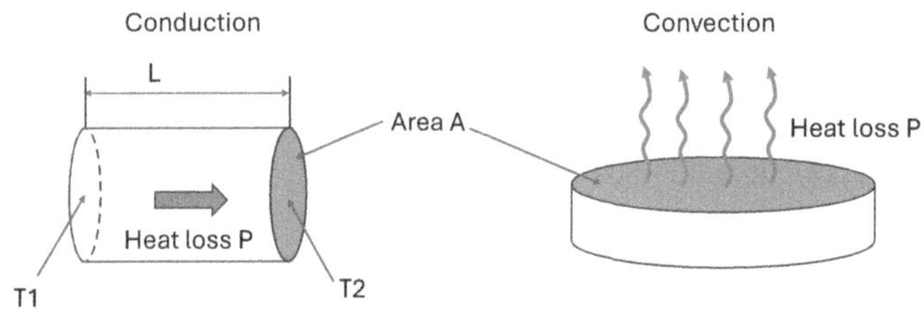

FIGURE 1.2 Illustration of conduction and convection

h: convective heat transfer coefficient (HTC)
A: convection surface area

In order to minimize thermal resistance for convection, one option is to increase the surface area A. The other is to increase the HTC h. The ways to increase h include:

1. Selecting fluid with higher h. Fluids with higher thermal conductivity, lower viscosity, and high specific heat capacity typically have higher HTCs. Liquid cooling can achieve higher h than air cooling. This explains why liquid cooling is the more efficient choice for EV traction system cooling.
2. Increase cooling fluid speed; use forced convection instead of natural convection.

In addition to different methods to improve cooling capability for EV traction systems, one critical element to achieve optimum thermal performance for electric drive systems is to minimize power loss and increase system efficiency. The physics is simple: for 100 kW power, 1% improvement in efficiency means 1 kW reduction in loss and a significant reduction in temperature rise and/or cooling system demands.

Those methods will be discussed extensively in the following chapters. Chapters 3 and 4 discuss traction motors and battery systems separately, where a large portion is dedicated to thermal management. Chapter 5 addresses power electronic systems and thermal design.

Chapter 2 will provide an overview of key vehicle performance characteristics including driving range, power, torque, acceleration, speed, charging, and efficiency. The chapter will describe how the electric drive system determines those characteristics.

Power electronic interconnect, the ‘weakest link’ in the power electronic system, will be discussed in Chapter 6. New developments to enable better performance and higher reliability will be included. Two key issues for the electric drive system, vibration, and electromagnetic interference (EMI), will be addressed in Chapter 7 and Chapter 8 separately. Several current major EV technology trends: modular approach, deep integration, SiC wide band gap (WBG) devices, and 800 V will be discussed in Chapter 9.

BIBLIOGRAPHY

Dhage, A. (2022, Dec 04). *Electric vehicle on board charger.* Retrieved from Battery Design from Chemistry to Pack https://www.batterydesign.net/on-board-chargers-ac-to-dc/

Ehsani, M. E. (2003). Characterization of electric motor drives for traction applications. *IECON'03. 29th Annual Conference of the IEEE Industrial Electronics Society (IEEE Cat. No.03CH37468).* Roanoke, VA, USA: IEEE.

Husain, I. E. (2021). Electric drive technology trends, challenges, and opportunities for future electric vehicles. *Proceedings of the IEEE, 109*(6), 1039–1059.

Kampl, S. (2020). *Topologies in on-board charging: Semiconductor recommendations for reliable charging solutions.* Retrieved from https://www.mouser.com/pdfDocs/Infineon-MOSFET_CoolMOS_CFD7A_topologies_in_OBC_-Whitepaper-v01_00-EN.pdf?srsltid=AfmBOoqBiFs8hBn1IGYrrELN2TymNOD6YkdKM5JOcGk5tVy3e4iPV3-O

Panasonic. (2023, Sep 22). *What is a DC/DC converter in an electric vehicle (EV)? - Device that converts high voltage into low voltage.* Retrieved from https://industrial.panasonic.com/ww/ss/technical/ap7

Rajashekara, K. (2013). Present status and future trends in electric vehicle propulsion technologies. *IEEE Journal of Emerging and Selected Topics in Power Electronics, 1*(1), 3–10.

Sgraham. (2021, Jul 01). *Why do EVs have a single speed transmission?* Retrieved from https://allev.info/2021/07/why-do-evs-have-a-single-speed-transmission/

US DoE Alternative Fuels Data Center. (n.d.). *Batteries for Electric Vehicles.* Retrieved from https://afdc.energy.gov/vehicles/electric-batteries

Yuan, J. (2021). A review of bidirectional on-board chargers for electric vehicles. *IEEE Access, 9*, 51501–51518.

2 EV Performance Characteristics and Electric Drive Systems

INTRODUCTION

According to IEA (2024), EVs have taken a significant market share from gasoline vehicles in a very short period of time, despite the disadvantages in driving range (energy density) and charging time. Low energy density is one of the primary disadvantages of EVs compared to gasoline vehicles. An EV battery has roughly 1/50–1/100 times energy density of gasoline by weight. EVs need bulkier and more costly battery systems to reach driving ranges to be competitive with gasoline vehicles. Charging time is another pain point. How can EVs come up with comparable or even better performance characteristics to win the market share? This chapter will attempt to provide some technical explanations.

First, EVs have significantly higher efficiency than gasoline vehicles, which makes up some disadvantages related to low energy density. They excel in instantaneous maximum torque delivery at zero speed. They also offer smooth, linear power delivery. As discussed in detail next, those characteristics are largely driven by their electric drive systems. Battery capacity determines driving range. Torque, power, acceleration, top speed, and efficiency rely on the motor, inverter, and battery. Charging time depends on chargers such as OBCs and DC fast chargers.

DRIVING RANGE

EV driving range, a key performance characteristic, mostly depends on its battery energy and vehicle weight. Average EV driving ranges are from 200 to 350 miles. The energy of an EV is characterized by its capacity, measured in watt-hours (Wh) or kilowatt-hours (kWh). They represent the amount of energy the battery can supply. For example, a battery with a capacity of 50 kWh can deliver 50 kW of power continuously in one hour. EV battery capacities typically range from 40 to 100 kWh (see Table 2.1). The higher the battery capacity, the more miles it can drive in one charge. Battery capacity can be estimated based on desired range and battery capacity per mile:

$$Battery\,Capacity\left(kWh\right) = Desired\,Range\left(Miles\right) \times Battery\,Capacity\,Per\,Mile \quad (2.1)$$

DOI: 10.1201/9781003569961-2

TABLE 2.1
Battery System Specification Comparison between Several Current EV Models (EV-Database, 2024; McDonald, 2018)

	Battery Capacity (kWh)	Range (mile)	Battery Capacity/ Mile (kWh/mile)	Battery Weight (kg)	Battery System Density (kWh/kg)
Audi E-Tron 55 quattro	95	225	0.422	~700	0.136
Porsche Taycan	89	300	0.297	~600	0.148
Tesla Model Y	60	215	0.279	~480	0.125
Nissan Leaf	40	145	0.276	~300	0.133
VW ID.4 GTX	82	260	0.315	~500	0.164
BMW IX XDrive 40	76.6	220	0.348	~500	0.153
BMW i3	33	114	0.289	N/A	N/A
Chevrolet Bolt	60	238	0.252	~435	0.138
Hyundai Loniq	28	124	0.226	N/A	N/A
Tesla Model 3	78	310	0.252	~544	0.143

Most of today's EVs use lithium-ion batteries. Higher efficiency leads to higher battery capacity per mile. Although these batteries may vary depending on manufacturing, battery capacity per mile is around 0.25–0.5 kWh/mile for many available EVs (see Table 2.1). Based on Lucid Motors (2022), for a standard range passenger EV in 2022, 0.25 kWh/mile can be achieved. This means a 200-mile EV needs a 50 kWh battery capacity. If the range increases to 350 miles, 70 kWh is needed. For example, the 2017 Tesla Model 3 base model is equipped with a 50 kWh battery capacity and has ~220 miles of range. This is equivalent to ~0.227 kWh/mile.

Other factors should be considered when determining battery capacity: battery degradation, battery efficiency changes with temperature and driving style, and battery pack loss.

In general, batteries lose 10–20% of range after 100,000 miles (Voelcker, 2024). Batteries lose capacity over time, and extra margins are typically needed to buffer this degradation. Battery packs are not 100% efficient due to factors like internal resistance and thermal management. Typically, capacity needs to be adjusted by assuming a loss factor of around 5–10%.

$$Adjusted\ Capacity = \frac{Required\ Capacity}{Pack\ Efficiency} \tag{2.2}$$

As discussed in Oreizi (2020), during actual application, factors such as wind, payload, tire traction, cold weather, HVAC, and road conditions affect energy loss and driving range. The energy usage (loss) for EVs can be grouped into four categories (Straubel, 2008):

1. Aerodynamic losses (drag from the air over the body of the car)
2. Tire losses (aerodynamic and rolling drag from the tires)
3. Drivetrain losses (battery, inverter, motor, transmission losses)
4. Ancillary losses (12 V loads, cooling fans and pumps, lights, etc.)

One critical factor related to drivetrain loss is driving conditions and speed. Driving at a speed greater than 65 mph reduces efficiency and range, since more power is needed at higher speeds; driving at 75 mph can result in a ~15% loss of range (Oreizi, 2020). Driving with frequent acceleration reduces battery range. Using the Tesla roadster as an example, under typical driving conditions, the power passing through the power electronic module is ~40 kW, and only 500 W of loss are generated by the inverter. However, under maximum power (218 kW for 30 seconds), there will be 6 kW of loss generated by the inverter, much higher than under typical driving conditions (Tesla, 2024).

Unlike gasoline-driven vehicles, battery performance varies with temperature. Under cold temperatures, the chemical reactions of lithium batteries slow down, and battery efficiency drops. EVs can lose up to 30% of their range in freezing temperatures (Sudbury, 2023).

To summarize, battery capacity is a main factor affecting the driving range of an EV. A higher capacity generally translates to a longer driving range on a single charge, but the actual range also depends on factors such as vehicle efficiency, driving conditions, and driving habits.

TORQUE

Torque is an important indicator of how much power an EV motor can produce. The higher the torque, the faster the acceleration, and the higher the vehicle towing capacity. As discussed earlier in Chapter 1, EV torque is generated by traction motors. It is mostly determined by the motor torque output, which is directly related to the flux density, current, and rotor volume. In EVs, where AC motors are commonly employed, torque is directly proportional to the AC current supplied by the inverter. Furthermore, a larger rotor volume generally signifies an increased capacity for handling higher levels of current and torque.

One key characteristic of EV torque is that electric motors can provide maximum torque at low or even zero speed when the electromotive force (EMF) is zero or very small. In comparison, for an ICE, maximum torque and power can only be achieved at high engine speeds. Electric motors react instantaneously as soon as the accelerator is pushed, with no need to reach high speeds. EVs exhibit superior torque and acceleration performance in comparison to ICE vehicles during acceleration (see Figure 2.1) (Kato, 2013; Whitehead, 2019).

There are two torque characteristics: peak torque and maximum continuous torque. Peak torque is the maximum torque that an electric drive system can deliver in a short period of time (up to 30 seconds). Peak torque is only available for a short duration. Prolonged periods of time risks overheating drivetrain components such as the inverter. In comparison, maximum continuous torque can be delivered

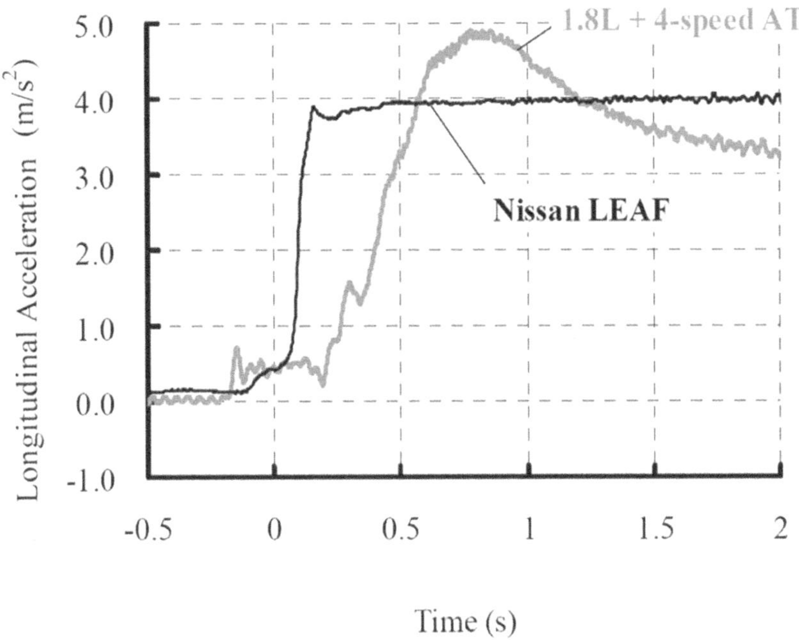

FIGURE 2.1 Acceleration response comparison between Nissan Leaf and an ICE vehicle (Sato, 2011).

continuously by the drivetrain system without overheating issues. Maximum continuous torque occurs during high-speed cruising on flat surfaces or when climbing uphill over a long period of time. Normally, the torque value provided by EV manufacturers is the peak torque (see Figure 2.2).

The reason why peak torque is immediately available at zero or low speed is that at an extremely low (or zero from a standing start) speed, there is little or zero back-EMF, which is proportional to BIV (B is flux density, I is current, V is speed). At low speeds, the back-EMF is negligible, and the supply voltage is equated to torque output when the throttle is fully engaged (shown in Figure 2.2). Consequently, peak torque becomes instantaneously available as maximum voltage is applied. It can be maintained as motor speed continues to increase. At the same time, peak power starts to increase and back-EMF opposing the supply voltage increases linearly with speed. Back-EMF reduces total voltage across the motor winding. As motor speed reaches approximately base speed, the back-EMF approaches the supply voltage and the total voltage across the motor winding approaches zero. Typically, the vehicle speed is around 35–45 mph at base speed (Munro Live, n.d.). Now the motor has reached peak power. As speed keeps increasing, back-EMF continues to increase alongside the speed; however, torque cannot be increased any further since there is no more voltage available to drive additional current through the windings. Peak power is maintained but peak torque begins dropping (Sieklucki, 2018; Hadraoui, 2022). For a typical EV, peak power is achieved at or around base speed (around

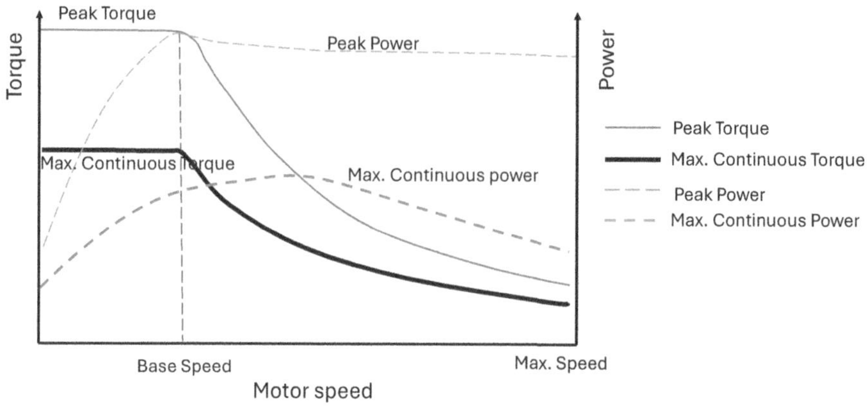

FIGURE 2.2 EV traction motor torque and speed characteristics regarding continuous conditions and peak condition (Shao, 2020).

35–45 mph for some vehicles). Using the Tesla Roadster as an example, the motor can produce a maximum torque of 375 Nm torque at motor speeds ranging from 0 to 4500 revolutions per minute (RPM) (base speed). As speed continues to increase, the torque starts to drop until speed reaches the maximum of 14,000 RPM (Tesla, 2024).

Peak torque of the vehicle is determined by the minimum of peak current from the battery, the motor, and the inverter. Similarly, the maximum continuous torque of the vehicle is dictated by the minimum of maximum continuous currents from the battery, the motor, and the inverter. Table 2.2 lists peak and maximum current limits for the first-generation Tesla Roadster battery, inverter, and motor. Currents for each component under specified conditions are generally close in a good design. For example, 640 A peak current is the drivetrain peak current that determines the vehicle's peak torque. It is only available for 10 seconds. On the other hand, maximum continuous torque is determined by 144 A maximum continuous current. Since peak current is much higher than maximum continuous current, peak torque is expected to be significantly higher than the maximum continuous torque.

Examples of EVs with high torque are listed here: Lucid Air: 1390 Nm, Tesla Model S Plaid: 1420 Nm, Porsche Taycan: 1050 Nm.

EVs experience maximum torque under the following driving conditions:

TABLE 2.2
Peak and Maximum Current limit for first Generation Tesla Roadster Electric Drive System (Tesla, 2024)

	Battery	Inverter	Motor
Maximum Continuous Current	144 A	175 Arms	144 Arms
Peak Current	650 A for 10 sec.	640 Arms	850 Arms

1. Maximum acceleration:
   As previously discussed, maximum acceleration happens when an EV accelerates from start, such as from 0 to 60 mph. The 0–60 mph performance refers to the time it takes for a vehicle to accelerate from zero speed to 60 mph. It is often considered a key indicator of a vehicle's ability to accelerate quickly. During 0–60 mph, the motor can deliver maximum torque to achieve maximum acceleration. Maximum acceleration is typically related to peak torque.
2. Towing or carrying maximum loads:
   When an EV is carrying a load as heavy as its maximum towing capacity, the motor may need to deliver maximum torque to overcome the resistance and transport the load for a long duration. This is related to maximum continuous torque.
3. Low speeds while driving uphill:
   Driving under low speed on a specified slope (grade) or hill hold, EVs can operate at, or close to, maximum continuous torque. When driving uphill, the motors may operate at, or close to, their maximum torque to provide the necessary torque as to overcome gravitational forces. One representative drive profile in EV powertrain design is 25/25; this is a combination of low-speed driving (25 mph) and climbing uphill (25% grade). It determines the torque needed for a vehicle to be able to be driven on a very steep hill (25% grade) at a minimum speed of 25 mph. Under these conditions, the vehicle can reach maximum continuous torque and maximum continuous current, and the inverter reaches one of the worst steady-state thermal conditions. Motors typically generate maximum torque at low speeds instead of high speeds.

   The gradeability is another parameter. Gradeability of an EV refers to its ability to climb a slope without losing speed or power. Electric motors with high torque capabilities, especially at low speeds, have better gradeability, allowing the vehicle to start or to maintain speed while climbing slopes.

POWER

Power of an EV is analogous to the horsepower of a gasoline vehicle. One horsepower is around 746 W. A higher power typically indicates a higher rate of work once moving, while torque is the capability of a car moving fast (high acceleration). When the power of an EV is specified, it typically refers to its peak power, the highest amount of power the engine can deliver in a short time.

As discussed earlier, power rating depends on how much power (voltage × current) the motor can handle and how much power the battery can provide through the inverter. Maximum power is related to both torque and speed, which is indicated by Equation 1.2. Similar to torque, there are two important metrics: maximum peak power and maximum continuous power. Maximum peak power of electric drivetrains is typically only available for short durations, such as during rapid acceleration or overcoming significant resistance (i.e., climbing a steep hill for a short period). Peak power is critical to ensure rapid acceleration and high-performance driving, allowing

EVs to quickly reach high speeds. The maximum continuous power enables the vehicle to reach and maintain a constant speed under stringent slope and speed conditions over an extended period without overheating motors and inverters. Maximum continuous power represents the sustainable power output of the EV under normal driving conditions, such as cruising at high speeds on freeways or carrying heavy loads over a long period of time.

One can picture peak and continuous power as the difference between sprinting 100 m and running a marathon. The 100 m sprint requires peak power just for a short time, while running a marathon needs continuous power for a very long time (Quora, 2020).

Maximum power is typically specified as peak power under transient conditions, while maximum continuous power means the maximum power at steady-state conditions. Both are critical conditions to consider when designing electric drive systems. These characteristics depend on corresponding parameters from the battery, the motor, and the inverter. Typically, traction motors have a higher peak power value than traction inverters due to their much larger thermal mass. Peak power for batteries is determined by the battery cells. Therefore, the peak power capabilities of the inverter and the battery together determine EV peak power capability. For the First Generation Tesla Roadster, its inverter can deliver 218 kW of power for 30 seconds, and its motor can deliver 218 kW of peak power for 10 seconds and 70 kW of maximum continuous power at 25°C ambient. The vehicle has a peak power of 215 kW, determined by its inverter and battery system capabilities (Tesla, 2024).

EVs can have anywhere from 50 to over 1000 hp (Moore, 2023), with the average being about 400 hp (294 kW). For example, the peak power of the Tesla Model 3 base model is ~211 kW (~283 hp), and the Nissan Leaf is about 147–214 hp (Car and Driver, 2024; Writer, 2024). As Thatcher (2024) discussed, the most powerful EVs on the market today are each equipped with two traction motors (see Table 2.3). Normally, the higher the power of the EV motor, the higher the torque it can produce.

Normal driving conditions do not need high power. Driving the Tesla Roadster typically only requires 40 kW despite the 215 kW of peak power it can deliver (Tesla, 2024). Economic EVs have lower horsepower. A lower-horsepower vehicle normally offers higher mileage. On the other hand, a higher horsepower rating usually means that the car can accelerate faster and has more torque available for climbing hills or passing other vehicles.

TABLE 2.3
EVs with High Power and Acceleration Performance (Lewis, 2023)

	Tesla Model S Plaid	Lucid Air Grand Touring	Porsche Taycan Turbo S	2022 Rivian R1S
Horsepower	1020	1050	750	1025
# of Motors	3	2	2	4
0–60 mph (seconds)	2.0–2.3	2.7	2.4	3.2

One peak power condition is 'wide open throttle' (WOT); it refers to the condition where the accelerating pedal of a vehicle is fully depressed, allowing the maximum amount of power available to be delivered to the electric motor. In most EVs, the electric motor also experiences the maximum torque available during WOT.

Traction motors need to maintain high power in a wide speed range in the constant power region, as shown in Figure 2.2. The continuous power region determines the maximum EV speed on flat surfaces.

ACCELERATION

Acceleration is the rate of change of velocity with respect to time. It is an important parameter to measure a car's performance. Again, the commonly used metric to measure acceleration is the time needed for the vehicle to start from 0 to 60 mph. The important factors affecting acceleration are torque output, aerodynamics, weight, tires, and assistance systems.

Higher power and force paired with lower mass (inertia) is always preferred to achieve more desirable acceleration. Based on Newton's second law, the acceleration of a vehicle is directly influenced by the force applied to it:

$$Acceleration = \frac{Force}{Mass} \tag{2.3}$$

Here, torque (T) is the rotational equivalent of force, acceleration is the angular acceleration (a), I is the moment of inertia.

$$\alpha = \frac{T}{I} \tag{2.4}$$

The acceleration of the vehicle equates to the product of angular acceleration (α) and wheel radius. Therefore, the ability of an EV to achieve maximum acceleration is closely tied to the electric motor's capability to deliver maximum torque instantaneously. In fact, maximum acceleration always comes with maximum torque. Equation 2.3 indicating weight of the vehicle is another critical factor that affects acceleration. A heavier vehicle means more inertia to accelerate.

Aerodynamics is important as well since the resistance due to drag rises quadratically with speed. Tires and roads are also critical to achieving high acceleration performance. Well-designed tires ensure optimum grip and minimal spinning during acceleration. Furthermore, electric drive systems will ensure maximum power output during acceleration.

Most all-electric cars can accelerate from 0 to 60 mph in less than 6 seconds (Kane, 2021). In comparison, typical gasoline vehicles need 6–12 seconds, while sports cars with gasoline can reach 4–6 seconds. Many high-performance EVs have 0–60 mph performance even below 2 seconds. Table 2.2 listed several sports EVs with high acceleration performance. They are equipped with high-power motors.

Electric motors offer a different driving experience with their quiet and responsive acceleration. This is reflected in the study (Hoshi, 2016), where the acceleration of an EV is compared to that of a gasoline vehicle. Gasoline vehicles overshoot their acceleration before stabilizing. In comparison, EVs have a shorter and smoother acceleration with minimal overshoot from start, which is largely because maximum torque is available right from a standstill.

TOP SPEED

Top speed is another important performance parameter. Most EVs have top speeds between 90 and 120 mph. High-power sports EVs can achieve higher speeds. For example, the top speed is 200 mph for the 2021 Tesla Model S; it is 168 mph for the 2022 Lucid Air, and around 155 mph for the 2021 Porsche Taycan 4S (Kane, 2021).

EV top speed is determined by its powertrain and battery characteristics, vehicle weight, aerodynamic design, and tire characteristics. As discussed in Chapter 1, the top speed of EVs is directly limited by the maximum AC frequency of the motor current, which is controlled by the inverter. The theoretical maximum speed of the vehicle is determined by its motor speed and transmission. Using the Tesla Roadster as an example, its maximum motor speed is 14,000 RPM and gear ratio is 8.27. Here, the tire diameter is 0.693 m.

First, the rotational speed of the wheels can be calculated by dividing the motor speed using the gear ratio.

$$\frac{Maximum\ Motor\ Speed}{Gear\ Ratio} = 1693\,RPM \tag{2.5}$$

Next, the top vehicle speed is determined by multiplying the theoretical maximum rotation speed of the wheels by the tire circumference.

$$Maximum\ rotational\ speed \times tire\ circumference = 137\,mph \tag{2.6}$$

This is in line with the maximum speed of the Roadster reported: 120 mph.

Driving at top speed is critical when evaluating electric traction system thermal performance at steady state. The power of the battery is another important factor in determining the top speed of an electric car, since a battery with more power will be able to provide more electricity to the motor.

Similar to an internal combustion vehicle, a car with a better aerodynamics design can be driven faster with less air drag. Lower drag coefficients and streamlined shapes contribute to reduced air resistance, allowing for higher top speeds.

In electric powertrain system design, trade-offs need to be considered between performance (driving range, acceleration, torque, and top speed) and efficiency. Quick acceleration is typically is associated with more torque, which leads to higher average speeds. However, this also corresponds to more loss due to increased air resistance, and higher motor loss. As a result, higher acceleration typically leads to lower efficiency and less overall driving range (Miles, 2017).

CHARGING

There are three levels of EV charging, which are differentiated by charging voltages (US Dept. of Transportation, 2023). Level 1 is typically used for residential purposes. Here, an EV is charged with the household 120Vac outlet. It carries current around 12–16 A. Power output will be ~1 kW. To charge an EV with 60 kWh, it will take 40–50 hours (Ramakrishnan, 2020). Level 2 is designated for commercial settings. Voltage is 208–240V and power output is around 7–19 kW. It will take around 4–10 hours to charge an EV with a 60 kWh battery capacity. It is apparent that for both Level 1 and Level 2, charging time is the bottleneck for EVs. Using an OBC (7.2 or 22 kW), it will take several hours to fully charge an EV with a 100 kWh capacity.

To increase power output and reduce charging time, either current or voltage needs to increase. Charging current is typically limited to ~400 A by the standard charging cable. Increasing voltage is a viable solution to increase charging power capability and reduce charging time. In fact, more and more EV manufacturers are introducing 800 V EV architecture to significantly reduce charging time, which will be discussed in Chapter 9.

One popular solution is Level 3 DC fast charging or supercharging under higher voltages (480–1000 V) (California Electric Vehicle Infrastructure Project, 2024). Here, power output can reach 50–350 kW, taking 20 minutes to 1 hour to fully charge a 60 kWh EV battery. Using the Tesla Model 3 (60 kWh) as an example (Electric Vehicle Database, 2023), with a 11 kW AC (OBC), it takes 6 hours and 15 minutes to charge the vehicle from zero to full capacity. Using fast charging (CCS), it takes 24 minutes to charge the vehicle from 10% to 80% capacity.

Power rating of OBCs determine on-board charging speed. Fast charging is undoubtedly the trend. However, due to the limited availability of DC fast chargers, AC OBCs that rely on the grid or home outlet are still the most popular solution. The success of on-board charging design lies in its capability to achieve efficiency and a compact design with high power density. Another ongoing trend is bidirectional charging, wherein the charger not only charges the vehicle, but also feeds the power back to the grid if needed.

EFFICIENCY

EVs are typically more than 60% efficient (Depart. of Energy, 2024), which is much higher than traditional gasoline vehicles. Only 20–30% of gasoline energy is actually used for driving. Therefore, the EV's high efficiency helps to mitigate its disadvantages of having low energy density.

Based on Depart. of Energy (2024), energy loss has two major sources:

1. Wheels: ~87% (including wind resistance loss, rolling resistance loss, and braking loss)
2. Electric drive system: ~18%

Here, the energy loss combined (105%) is higher than 100%, because regenerate braking recaptures energy and reuses it.

Efficiency is affected by several factors:

1. Drivetrain efficiency. One of the major sources is drivetrain efficiency. Here, drivetrain includes motors, inverters, and transmissions. The efficiency varies with driving conditions. As discussed earlier, efficiency of motors is higher than 90% under typical driving conditions, reaching 95% for optimal designs. EV traction inverters operate with efficiencies above 90%, often reaching into the 95–98% range.
2. Battery efficiency. Lithium-ion batteries can achieve efficiencies of 90% at the beginning of their life. It can reduce to 75% at the end of the life (Koroma, 2022).
3. Regenerative braking. Regenerative braking systems are implemented to improve EV efficiency. They capture and convert some of the kinetic energy from braking back into electrical energy. The electrical energy, which would have been lost, is now used to recharge the battery. Thus, EV efficiency is around 60–73% but can exceed 77% with regenerative braking.
4. Vehicle weight and aerodynamics. Heavier EV loads require more driving energy, resulting in less drivetrain efficiency. At the same time, heavier vehicles increase tire resistance, resulting in additional losses. Similar to the ICE, optimal vehicle aerodynamics lead to less air drag and less energy loss. Higher speeds introduce higher losses since air resistance increases exponentially with speed. At high speeds, aerodynamic drag for EVs increases with square of the speed. If the speed increases from 40 to 80 mph, the loss due to aerodynamic drag increases 4 times.
5. Driving conditions. Vehicle efficiency varies with driving conditions. One major reason is that the efficiency of the drivetrain, including traction motor and inverter, varies with driving conditions. Drivetrain efficiency under city driving is different from that under steady highway driving. Driving in cold weather also impacts battery performance and reduces efficiency.
6. Charging efficiency. Charging efficiency is the ratio of the energy stored in the battery during charging to the energy provided by the grid. Charging efficiency varies with the charging method. Level 1 and Level 2 charging efficiency mostly depends on OBC efficiency. OBCs can achieve efficiencies above 90%, often in the range of 92–95%. Efficiency around 85% has been reported for Level 1 and Level 2. Level 3 charging can achieve higher efficiency, with ~90% reported (Nedelea, 2024).
7. Accessory and auxiliary electrical losses. The use of accessories such as air conditioning, lights, and climate control systems is considered an additional load that increases power losses and reduces vehicle efficiency.

An electric drive system, including the drivetrain, battery, and regenerative braking, is only responsible for a 15–20% energy loss. Compared to 64–75% for an ICE, the electric drive system is much more efficient. Motors, inverters, and DC-DC converters typically have efficiencies well above 90%, in the mid to high 90s. This is the main reason for EVs being significantly more efficient than traditional gas vehicles.

EV efficiency is heavily influenced by the capabilities, designs, and operations of its electric drive system, especially with regard to its traction motors, inverters, and battery systems. The more efficient the electric drive system, the more battery power is converted into mechanical power, which results in a longer driving range. In addition, most of the energy losses produced in electric drive systems are heat losses. A significant portion occurs due to joule heating, switching loss in the semiconductor switches, and magnetic component losses. Those losses need to be dissipated effectively to prevent components from overheating and to maintain reliability and performance. The more efficient the electric drive system, the less heat will be generated, which leads to either less cooling system needs or lower operating temperatures. Lower cooling needs mean potentially both weight and cost savings, and lower operating temperatures increase reliability. These will be discussed in greater detail in the next chapters.

BIBLIOGRAPHY

California Electric Vehicle Infrastructure Project. (2024). *EV charging basics.* Retrieved from California Energy Commission https://calevip.org/electric-vehicle-charging-101

Car and Driver. (2024). *2025 Porsche Taycan.* Retrieved from Car and Driver https://www.caranddriver.com/porsche/taycan

Depart. of Energy. (2024). *Where the energy goes: Electric cars.* Retrieved from https://fueleconomy.gov/feg/atv-ev.shtml

Electric Vehicle Database. (2023, Sep). *Tesla model 3.* Retrieved from https://ev-database.org/car/1991/Tesla-Model-3

EV-Database. (2024). Retrieved from Electric Vehicle Database https://ev-database.org/#sort:path~type~order=.rank~number~desc|rs-price:prev~next=10000~100000|rs-range:prev~next=0~1000|rs-fastcharge:prev~next=0~1500|rs-acceleration:prev~next=2~23|rs-topspeed:prev~next=110~350|rs-battery:prev~next=10~200|rs-towweight

Hadraoui, E. A. (2022). A multi-criteria analysis and trends of electric motors for electric vehicles. *World Electric Vehicle Journal, 13*(4), 65. Retrieved from https://www.mdpi.com/2032-6653/13/4/65

Hoshi, M. E. (2016). Electric vehicles and expectations for wide bandgap power devices. *28th International Symposium on Power Semiconductor Devices and ICs (ISPSD).* Prague, Czech Republic: IEEE.

IEA. (2024). *Global EV outlook 2024.* Retrieved from https://www.iea.org/reports/global-ev-outlook-2024

Kane, M. (2021, Sep 23). *Electric cars listed by 0-60 MPH acceleration: Quickest to slowest.* Retrieved from https://insideevs.com/news/534101/electric-cars-60mph-acceleration-20210919/

Kato, T. E. (2013). Advanced technologies for traction motor for automobile. *2013 IEEE ECCE Asia Downunder.* Melbourne, VIC, Australia: IEEE.

Koroma, M. E. (2022). *Life cycle assessment of battery electric vehicles: Implications of future electricity mix and different battery end-of-life management.* Retrieved from https://pmc.ncbi.nlm.nih.gov/articles/PMC9171403/

3 EV Traction Motor and Thermal Design

INTRODUCTION

Many EVs and HEVs use PMSMs. The list is long: Toyota Prius, Chevrolet Bolt EV, Ford Focus Electric, Nissan Leaf, Honda Accord, BMW i3 and i8, Jaguar I-PACE, and some Tesla models. They accounted for more than 75% of AC traction motors in battery electric vehicles (BEVs) and HEVs in 2023 (Gobbi, 2024; Edmondson, 2022), while induction motors accounted for 17%. PMSMs are the choice for EV traction motors due to their high efficiency and high power density.

Current EV traction motors can maintain constant high torque at low speeds and constant high power at higher speeds, making them suitable for traction applications where both high torque during acceleration and high-speed driving capability are needed. On the other hand, traction motors generate a significant amount of heat. The heat is mostly generated by windings and cores. Traction motor cooling needs to consider the motor heat loss and the thermal resistance of the motor structures.

PERMANENT MAGNETIC SYNCHRONOUS MOTOR

A PMSM (shown in Figure 3.1 and Figure 3.2) has two major parts: stator and rotor. The stator is the stationary part, and the rotor is the rotating part. The rotor has permanent magnetics that generates a magnetic field with constant flux. The stator includes windings through which AC passes. The winding current will generate a magnetic field. When the motor is running, the magnetic field from the stator rotates due to the AC. The magnetic field created by the rotor magnetics endeavors to align with this rotating magnetic field, resulting in the rotor spinning.

Typically, the PMSM stator is located on the outside of the motor, and rotor is on the inside of the motor. The stator of the PMSM includes an out frame and a stator core with windings. So far, PMSMs for EV traction applications use three phases. The stator has three sets of windings, and the inverter supplies three phases of AC respectively. The inverter precisely controls the amplitude and phase of AC, which determines the magnetic field strength and direction. The winding is wound in a special pattern around the stator core in order to produce the desired magnetic field. The rotor includes permanent magnets and a rotor core. The magnetics are fixed to the rotor core.

Pole number is one critical motor design parameter. Every magnetic has at least two poles, North and South. The PMSM rotor magnets create an even number of magnetic poles. The number of magnetic poles in the rotor decide the pole number

 DOI: 10.1201/9781003569961-3

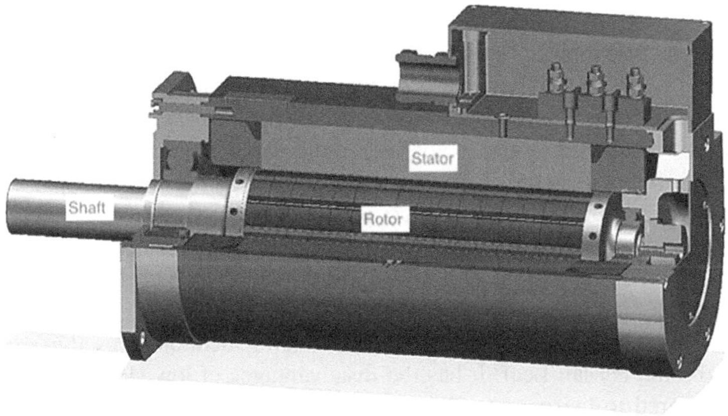

FIGURE 3.1 Illustration of PMSM (Egzon123, Wikimedia Commons, licensed under Creative Commons Attribution-Share Alike 3.0 Unported license, from https://commons.wikimedia.org/wiki/File:-synchron-motor-.jpg).

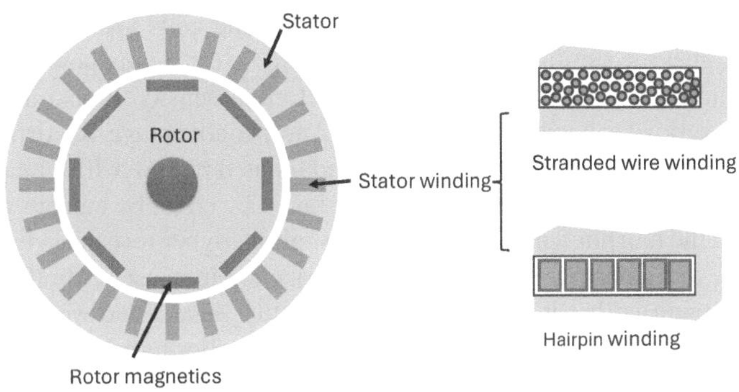

FIGURE 3.2 Illustration of stranded wire winding vs. hairpin winding in one slot for motors.

of the motor. Eight poles and twelve poles are typical for EV traction motors (Krings, 2020). As Chapter 1 revealed (see Equation 1.4), the motor's pole number affects the motor's speed. In the permanent magnet motor, the magnet is either embedded in or affixed to the rotor surface. Permanent magnetics in the rotors contribute to the flux density. It can achieve high power density, high efficiency, and high starting torque. Those characteristics make the PMSM a popular choice for EV traction motors. However, its cost is high due to the use of permanent magnetics, which requires rare earth materials and have high production costs.

The motor shaft is the rotating part that rotates with rotor. It provides support to the rotor. The shaft is typically made from high-strength materials such as steel or alloys. The shaft is supported by bearings at both ends. Motor bearing is another critical element to ensure the shaft spins with minimal friction.

Stator winding

Winding is a critical piece in the motors. It is normally made of copper with insulation around it. EV traction motors have two types of winding technology: stranded winding and hairpin winding (bar winding) (See Figure 3.2):

Table 3.1 compared stranded winding and hairpin winding. One aspect is slot fill, which is the percentage of the slot area occupied by the stator copper windings. Low slot fill corresponds to a smaller copper cross-section area, which leads to lower power density and higher resistance. Another aspect is end turns, the portion of the winding that goes around the end of the stator tooth. The end turn adds resistance length and does not help performance.

Stranded winding is the traditional motor winding technology used in some Tesla models and the Nissan Leaf. It has the disadvantages of low slot fill and long end turns compared to hairpin winding. Stranded winding has higher electrical resistance and more winding losses.

The hairpin winding configuration is relatively new to the industry and has become a more popular choice for EV traction motors. Hyundai Electric Global Modular Platform (E-GMP) and Chevrolet Volt/Bolt use hairpin winding. Hairpin windings are more tightly wound with rectangular cross-sections than stranded windings with circular cross-sections. The tighter the coils are wound around the stator, the higher the slot-filling factor is, and the more efficiently power is generated. Hairpin winding allows for higher power and torque density. It also improves efficiency and overload capacity (Husain, 2020). Another advantage of hairpin windings is their excellent thermal performance. Hyundai has reported a 10% increase in the coil space factor (the ratio of available space taken by coils) by hairpin winding. At high speeds, the hairpin winding may have a slightly higher resistance than stranded winding due to higher AC resistance when frequency increases. Another limitation of hairpin design is that it has much less flexibility than round-wound winding at the design stage (Quassolo, 2022; Husain, 2020). Smaller hairpins are emerging as a popular choice to take advantage of both stranded and traditional hairpin designs. Tesla has used 2 mm wide hairpin windings.

Winding loss is a major part of the losses generated in PMSMs. Motor design needs to minimize winding loss and dissipate heat loss effectively.

TABLE 3.1
Comparison Between Stranded Winding and Hairpin Winding (Munro Live, 2024; Krings, 2020)

	Stranded	Hairpin
Resistance	AC resistance ≈ DC resistance	AC resistance>DC resistance
Slot Fill	Low (0.4–0.5)	High (>0.7)
End Turns	Long	Short
Connection	Traditional	Need connection ring

There are two types of stator winding methods: distributed winding and concentrated winding (see Figure 3.3). In concentrated winding, only one winding is wound around one tooth of the stator tooth. In distributed winding, at least two windings are wound around one tooth. Distributed winding typically leads to a smoother sinusoidal back-EMF and lower harmonics than concentrated winding. Therefore, stator core loss and winding losses will be smaller. This is very important for traction inverters where high efficiency is critical. Most current EVs use distributed winding (Kano, 2013; Husain, 2020). The examples include Tesla Model 3 (2017) and Model S 60 (2013), BMW i3 (2016), Nissan Leaf (2012), and Chevrolet Bolt (2017).

In terms of thermal management, most winding loss must be dissipated through conduction, through either the wires' length or cross-section. The insulations have very low thermal conductivity, and air gaps around insulations are pretty common. Both make heat dissipation difficult. A higher number of wires will result in higher thermal resistance due to the presence of insulation layers and air gaps. This makes heat dissipation even more challenging. This is the reason that hairpin winding has better thermal dissipation capability.

Rotor Magnetics

Rotor magnetics is a set of magnets located either internally or externally on the rotor core. If the magnetic is located externally, it is called a surface permanent magnet (SPM) synchronous motor. If the magnet is embedded internally, it is called an interior permanent magnet (IPM) synchronous motor (see Figure 3.4). Most current EVs use IPMs, including the Tesla Model 3, BMW iX3, Nissan Leaf, Chevy Bolt, and Toyota Prius (Husain, 2020; Lammotor, 2024). With a high magnetic saliency ratio, IPMs can take advantage of reluctance torque and offer higher power/torque density. IPMs have advantages for high-speed applications (Kato, 2013; Vector Magnets, n.d.).

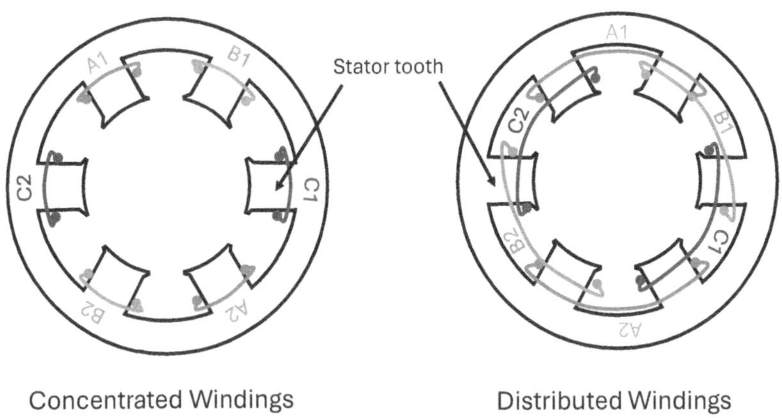

FIGURE 3.3 Comparison of concentrated winding and distributed winding in the stator.

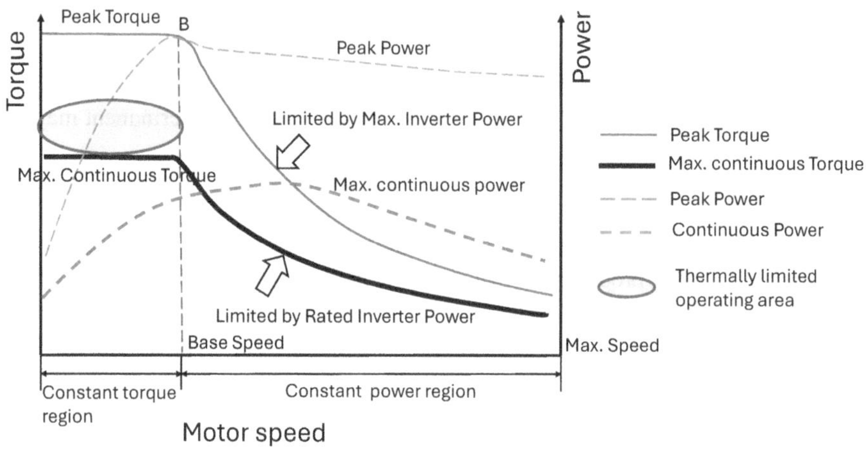

FIGURE 3.5 Torque, speed, and power characteristics of an EV traction motor (Roy, Feb 2021; Shepard, 2024; Sha, 2020; Gobbi, 2024).

Maximum continuous torque is limited by motor thermal performance. As indicated in Figure 3.5 (Roy, Feb 2021), the purple circle corresponds to a high torque area limited by thermal performance. In this area, the motor will draw maximum continuous current. In actual design, 70% or higher of maximum continuous torque is typically selected as the maximum continuous motor torque, considering design margin.

As Hughes (2013) states, torque is proportional to rotor volume, AC current, and flux density (*B*) for given values of the specific magnetic and electric loadings.

$$T = \mathrm{k}\frac{\pi}{4}\left(\underline{B} * \underline{A}\right)D^2 L \tag{3.1}$$

B: radial flux density
A: axial current per meter of circumference on the rotor
$\pi D^2 L/4$: rotor volume (*D*: rotor diameter; *L*: rotor length)

First, torque is typically proportional to its rotor size. The larger the rotor size, the higher the torque. Akatsu (2013) listed torque volume density for several released EV/HEV permanent motors, showing that torque is roughly proportional to motor size under the same type of cooling and powertrain system.

Secondly, higher flux density (*B*) leads to higher torque. There are several options to increase flux density. One option is to reduce the air gap between the stator and the rotor. This is the reason the gap between the rotor and stator is designed to be as small as possible. The second option is to increase the current flowing through the stator winding. For traction motors, its flux density is directly related to the winding current. Peak torque is related to peak current, and maximum continuous torque is related to maximum continuous current. As shown in Table 2.1, for the Tesla

Roadster traction motor, peak current (850 Arms) is much higher than maximum continuous current (144 Arms). Higher winding current produces higher torque. However, the winding current is limited by thermal design. Heat loss in the winding due to joule heating (I^2R) increases with the square of the current. Better cooling means more heat can be dissipated, and more current can be applied without exceeding the temperature limit. Therefore, torque capacity also depends on the cooling design (Mohan, 2002).

There is a maximum value of flux density a motor can have without introducing saturation in the laminations. Motor efficiency, motor torque output, and motor power output will drop significantly if saturation occurs. PMSM laminations typically have a high saturation flux density range from around 1.5 to 2.0 T. This allows the motor to handle high magnetic flux without saturating, which enables high torque density.

Maximum continuous torque which a motor can supply mainly depends on its thermal characteristics. At the same time, it takes a while for the motor to reach its steady state due to its large thermal mass. Therefore, the peak torque for a traction motor is normally much higher than its maximum continuous torque (see Figure 3.5) since duration of the peak torque (less than 1 minute) is small compared with the thermal time constant of the motor. Typically, the motor needs to provide high torque twice that of the maximum continuous torque (rated torque) for a short period of time (15–20 seconds). Thus, the peak torque of a traction motor is normally twice that of its maximum continuous torque (Shao, 2020).

The limiting factor for peak torque (current) of an electric drive system can be either the battery or inverter. Inverter current is most likely limited by the maximum junction temperature of the power devices. The thermal time constant of the inverter is much smaller than that of the motors. It takes much less time for the inverter to reach peak current conditions than the motor. On the other hand, the amount of current that the battery system can supply to the motor also affects peak torque. This will be discussed in the next chapter. (The thermal time constant is the time taken by a mass such as the inverter or motor to reach ~63% of its steady-state temperature while heating up or cooling down. It is calculated by multiplying the thermal capacitance of the mass C_{th} with the thermal resistance R_{th}. It takes a much longer time for the motor to reach steady state than for the inverter, due to the inverter's smaller mass.)

For EV motor design, it is desired to have high torque at start or low speed for acceleration and touring performance. In addition, sufficient power at high speed (cruise) and a wide speed range is desired (Yildirim et al., 2014).

Power is to quantify the traction motor's ability to do work over time. Traction motor performance includes peak power and maximum continuous power. Both are important. Figure 3.5 is an overview of EV traction inverter power, torque, and speed characteristics. It includes both peak and continuous power performance. Peak power is related to acceleration or other high-demanding conditions such as climbing a steep slope over a short time duration, typically from a couple of seconds to a few minutes without overheating. When it exceeds this short time duration, the motor or inverter temperature will rise beyond the temperature limit of the material, causing failures such as insulation breakdown or demagnetization of the magnets. On the other hand, maximum continuous power is the power that the motor can deliver continuously as long as cooling is maintained. The traction motor needs high continuous power during heavy load driving or steep grade climbing. Those two critical conditions need to be considered in traction motor design. Shao (2020) pointed out that the motor needs to be able to provide consistent power at higher speeds for cruising of three to four times the base speed. This is another critical condition to be considered. Using the Tesla Roadster as an example, its motor can deliver 218 kW of peak power for 10 seconds and 70 kW of maximum continuous power at 25°C ambient (Tesla, 2024).

The power and torque ratings of EV traction motors listed by EV manufacturers are mostly peak power and peak torque. Table 3.2 lists several EVs' critical motor characteristics. These include peak power, peak torque, top motor speed, power density, and maximum motor efficiency. As we can see, generally, an EV with higher peak power also has higher peak torque.

Speed

As discussed in Chapter 2 (see Equation 1.4), traction motor speed is precisely controlled by the inverter adjusting AC frequency. Maximum motor speed is limited by maximum AC frequency, which is typically limited to ~1.5 kHz (Quassolo, 2022). Husain (2020) provided the AC frequency value for several EVs. The Tesla Model 3 (2017) is 900 Hz. The Nissan Leaf (2012) is 693 Hz. Since motor output power is the product of torque and rotational speed, increasing motor speed can achieve higher power output without increasing torque or rotor volume. In fact, increasing rotor speed is the main approach to achieve higher power density for EV traction motors over the last decades (Krings, 2020).

Maximum motor speed is also limited by rotor strength and bearing life. Rotor laminations need to have sufficient strength to withstand the high centrifugal force due to high speed. In order to spin at higher speeds (>20,000 RPM) without severe rotor lamination deformation or even damage, the Tesla Model S Plaid uses a

TABLE 3.2
Parameters for Several EV Traction Motors (Shao, 2020; Husain, 2020; MotorXP-PM , 2020)

	Peak Torque (Nm)	Peak Power (kW)	Top Speed (RPM)	Power Density (kW/L)	Maximum Efficiency (%)	Winding	Cooling
Nissan Leaf (2012)	280	80	10,390	4.2	97	Distributed	Water
BMWi3 (2016)	250	125	11,400	9.1	94	Distributed	Water
Tesla Model 3 (2017)	410	192	18,000	N/A	~95	Distributed	Oil
Chevy Bolt (2017)	360	150	8810	N/A	N/A	Distributed	Oil

carbon-fiber sleeve wrapped around the rotor to increase its mechanical and electromagnetic performance. The tensile strength of carbon fiber (>3000 MPa) is much higher than that of steel (~1500 MPa). In addition, the carbon fiber sleeve allows for a smaller air gap between the rotor and the stator. This increases the magnetic flux density and efficiency. Traction motors usually use deep groove ball bearings, where the cage, lubrication, and cooling are limiting factors. As rotor speed increases, the bearing will be under high load, and its lubrication and cooling will become more difficult. These factors limit the maximum rotor speed. Another important factor to consider is the thermal limit, because the core loss and winding AC losses also increase with speed (AC frequency).

Traction motors will operate over a wide range of RPMs with high efficiency. This requires traction motors to operate at higher RPMs than combustion engines. Chapter 2 explained how to determine EV top speed from motor speed. Table 3.2 listed maximum motor speed for several typical EVs. 'Maximum speeds for traction machines found in the industry are in the range of 12–20 KRPM. The trend is to move towards maximum speeds of the traction motor above 20 KRPM, in particular for high-performance electric vehicles' (Quassolo, 2022; Shao, 2020). The Chevrolet Bolt used a traction motor with a maximum speed of 8810 RPM. The Nissan Leaf used an EM57 traction motor with a maximum speed of 10,390 RPM.

Efficiency

The motor needs to provide high efficiency over the entire speed range. Efficiency is critical as it affects the vehicle's performance, driving range, and cooling system load. The efficiency of a traction motor is its mechanical power output divided by the total amount of power input (see Figure 3.6). Efficiency indicates how effectively the motor converts electrical power to mechanical power. The difference between electrical input power and mechanical output power is the loss. During the energy

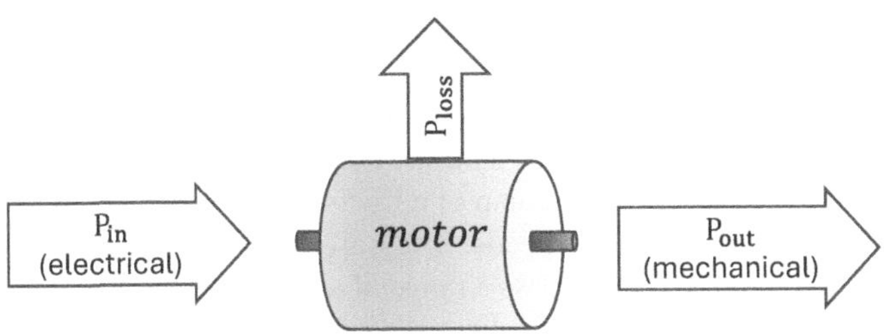

$$P_{loss} = P_{in} - P_{out}$$

$$efficiecy = P_{out} / P_{in} = 1 - P_{loss} / P_{in}$$

FIGURE 3.6 Illustration of power flow in motors.

conversion process, power loss is converted to heat. Higher efficiency means less heat.

The mechanical power output is always less than the electrical power input due to various losses.

As stated in Jenkins (2023), 'Approximate drivetrain efficiency ranges from 75–90% for the EV to a rather less impressive 20–35% for the ICE vehicle,…the main drivetrain components in an EV battery, inverter, motor, and gear reduction and/or transmission – all have efficiencies in the mid-to-high 90% range. Out of the main drivetrain components in an EV, most of the losses come from the motor.' Typically, EV motor efficiency varies with types of motor and driving conditions. 90% or higher efficiency is expected for most frequent driving conditions. 95% has been reported (See Table 3.2).

PMSMs are more efficient than induction motors. The maximum efficiency of induction motors is about 90–93%, while maximum efficiency of EV PMSMs can reach 97% (Hanejko, 2022; Jenkins, 2023). The reason is that the induction motor needs additional power to create an electromagnetic field, reducing its efficiency at high torque. PMSMs have an efficiency of ~2% more than a highly efficient induction electric motor, provided that the stator has the same design and the same variable frequency drive is used for control (About-Motors, 2023). 'The use of permanent magnet synchronous motors enables (Tesla) Model 3 to match the 100 kWh (Tesla) Model S 100D with a battery capacity of 75 kWh' (Marchio, 2024). The efficiency advantage for PMSMs is more obvious at low-speed conditions (Munro Live, n.d.). But 'Induction machines have perfect flux regulation and therefore you can optimize your efficiency' stated Konstantinos Laskaris, Tesla's Chief Motor Designer (EVANNEX, 2018). It is noted induction motors have better efficiency at high speeds. This is the reason that some high-performance EVs are equipped with two motors: one induction motor for high-speed driving conditions and one PMSM for low to medium-speed driving condition.

Efficiency depends on the driving conditions. Driving under maximum torque or high speed is less efficient, and this can be illustrated with an efficiency map. The efficiency map of a traction motor is a graphic representation indicating how efficiently a motor operates under various conditions, as a function of two key performance factors: torque and speed (Jenkins, 2023).

Figure 3.7 illustrates efficiency map of a traction motor. It plots motor efficiency values at different torques and speeds. They are created by testing, where motor input and output are measured across a range of speeds and torques.

There is a high efficiency area where the motor has minimal loss. It typically happens at around base speed for permanent motors. A low efficiency area normally appears at low torque or low speed. For example, the 2012 Nissan LEAF motor has peak efficiency of about 97% between 5000 and 9000 RPM. Under most conditions, efficiency is higher than 90%. The lowest efficiency is less than 70%, which corresponds to very low speed and/or very low torque (Burress, 2013).

Several important regions in the motor efficiency map are addressed here (shown in Figure 3.6):

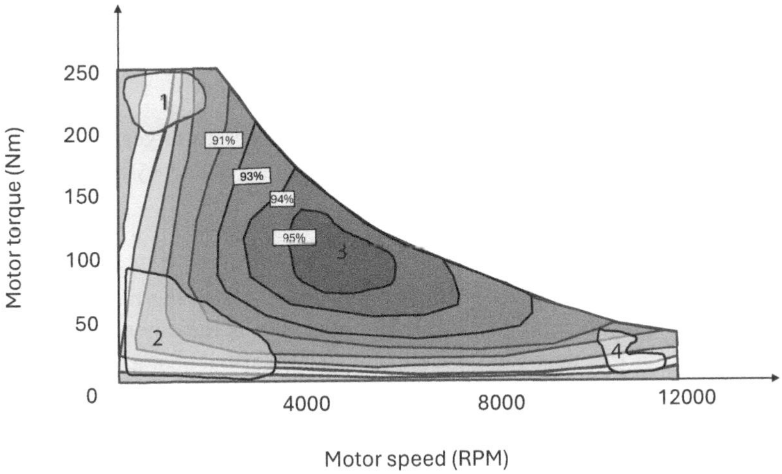

FIGURE 3.7 Efficiency map of a traction motor.

Region 1: High Torque Region (Area 1 in Figure 3.7): This region corresponds to low speed and high torque. It occurs when the vehicle starts or is driven at low speed and peak torque (maximum current) is applied. Typical conditions include WOT, touring with a heavy load, and driving uphill on a steep grade. Using WOT as an example, when the vehicle starts from zero RPM, maximum acceleration is commanded, which corresponds to peak torque and peak current. There will be a significant amount of joule loss (I^2R) due to peak current, largely occurring in the windings. On the other hand, since power output is proportional to torque times speed (RPM) and the speed is low, the total power applied is very low. Therefore, efficiency is low and loss is high. Both worst-case scenarios (transient thermal and steady-state thermal) occur in this region.

Region 2: Low Speed and Low Torque (Area 2 in Figure 3.7): In this region, the power output to drive the vehicle is low due to low speed and low torque. Although the losses related to torque and speed are low, the other losses such as winding loss, core loss, and mechanical friction loss are significant. As a result, losses are becoming a large portion of the total input power, and the efficiency is low.

Region 3: High Efficiency Region (Area 3 in Figure 3.7): Region 3 is the optimum region where efficiency is the highest. It typically occurs when the load and torque are around base speed. According to (Munro Live, n.d.), permanent magnets will reach high efficiency in lower speed regions such as city driving, while the high efficiency region for induction motors occurs at higher speeds such as highway driving.

Region 4: High Speed and Low Torque Region (Area 4 in Figure 3.7): Motor loss includes mechanical loss and electrical loss such as eddy current and hysteresis loss. Both mechanical loss and electrical loss increase significantly with motor speed. Therefore, motor efficiency will be lower at high speed for permanent magnet motors.

In order to achieve high efficiency, motor design needs to target the highest efficiency at desired driving conditions based on the motor efficiency map.

THERMAL MANAGEMENT

Thermal management of motors includes three aspects: loss, thermal characteristics of the motors structure, and cooling design. The loss is the amount of heat generated by the motor that needs to be dissipated. The thermal characteristics of motor structure determine how difficult it is to dissipate the heat through the motor without active cooling. Cooling design involves how to design active cooling to effectively dissipate the heat.

Thermal management for EV traction motors is challenging for two reasons. First, traction motors are designed to handle high power density: compact and lightweight while delivering the highest power output. High power density leads to high power loss density. It is always more difficult to handle higher power loss density. Secondly, the motor structure lacks an effective heat dissipation path. The wide presence of low thermal conductivity layers such as insulations and air gaps make it difficult to dissipate heat effectively.

MOTOR LOSS

Loss dictates thermal design of the traction motor. Heat generated by the motor power loss results in a temperature rise. Motor loss contributes to the highest loss in the EV traction system. As discussed earlier, motor efficiency varies under different driving conditions with respect to torque and speed. Driving conditions largely determine the loss. Driving under high torque or high-speed conditions will result in a dramatic reduction of efficiency and an increase in loss.

The loss in a motor typically includes mechanical loss and electrical loss (Roshandel, 2021) (Figure 3.8). Mechanical loss consists of friction loss and windage loss. Electrical loss includes winding loss and core loss.

Friction losses are related to the resistance against the rotor's rotation. They are typically due to the friction of the bearing and increase with rotor speed. Windage loss occurs due to the movement of air or oil to resist the rotor from spinning. Windage losses are proportional to the cube of rotor speed. Therefore, high-speed driving will result in a significant increase in windage loss (Jenkins, 2019).

Winding loss (I^2R) is largely determined by the winding current and the winding resistance. Since winding resistance changes with frequency due to skin effect and proximity effects, the winding loss is also a function of frequency for traction

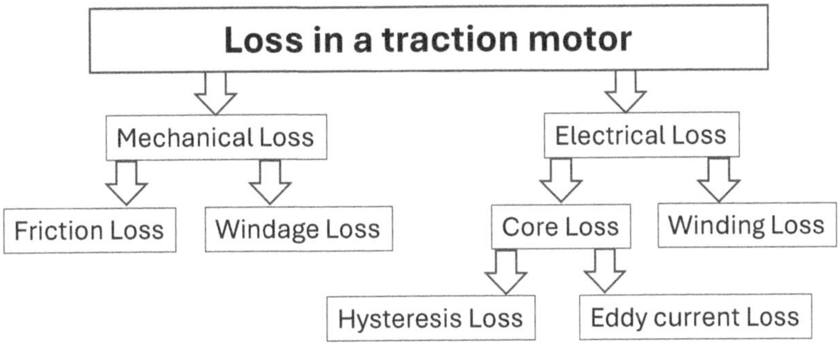

FIGURE 3.8 Loss in a traction motor.

motors, which are normally operated under AC. The skin effect forces AC current flow towards the surface of the winding wires and results in additional resistive loss (Jenkins, 2019).

Under high torque conditions, the winding loss becomes significant as the loss increases with square of current. In traction motors, winding current is directly related to the motor torque. The motor speed is directly related to the AC current supply frequency. As a result, winding loss is affected by both torque and speed of the motor (Roshandel, 2021).

Winding loss is a large part of the total motor loss and is converted into heat loss. Motor windings are made of copper. For EV traction motors, stator winding loss tends to be a dominant portion of the total motor loss. For induction motors, winding loss also happens at the rotor bar.

Core loss, also called iron loss, occurs in the permanent magnetics and laminations. It includes hysteresis loss and eddy current loss. Core loss increases with frequency and leakage flux. Hysteresis loss comes from magnetic materials and laminations. When the magnetic field changes, magnetic material and lamination material have to respond. They go through magnetization and demagnetization to adjust their magnetic orientations. This causes hysteresis loss. The loss depends on the amplitude and frequency of the magnetic field. The faster the magnetic field changes, the higher the loss (OSWOS, 2024). Eddy current loss is joule heating loss due to eddy currents. When the magnetic field changes, it will induce an EMF opposing the change, which generates eddy currents in the lamination. The amount of loss is related to square of the current. As discussed earlier, one typical method to reduce eddy current loss is to use laminations that are electrically insulated from each other. This increases the core's resistance and reduces induced EMF. The thinner the lamination, the lower the eddy current loss. As Peng (2018) has shown, core loss in permanent magnetic motors can be estimated via the Bertotti model, where the loss is proportional to the frequency of the alternating magnetic field and magnetic flux density.

To summarize, motor loss is directly proportional to speed and current. Higher torque and higher speed lead to higher loss. The increase in winding loss increase is related to the square of the current.

Thermal Resistance

The motor is made of winding, magnetics, lamination, and insulations. The temperature limit of traction motors is determined by both insulations and the magnetic material used. Windings and laminations are metals with a high temperature limit. NdFeB, the most popular magnetic choice for EV traction motor, has good demagnetization performance when the temperature is below around 150°C for some cases. Typically, electrical insulation is made of organic polymers. The insulation can degrade at high temperatures or under high thermal stress as a result of large temperature changes when the current is switched on and off. This can eventually lead to insulation damage and motor failure. The maximum temperature for most motors' insulation is often around 105–200°C (Konovalov, 2023).

Motor structure lacks an effective heat dissipation path and results in high overall thermal resistance. In motors, a large portion of heat loss occurs at the stator winding. The other source is magnetic losses from the laminations. Both are surrounded by layers of low thermal conductivity materials such as insulations and air gaps. Thermal properties for typical motor components are provided in Table 3.3.

One study (Wang, 2023) shows there is a significant motor portion with low equivalent thermal conductivity in both radial and axial directions. This contributes to low thermal dissipation capability in both directions (see Figure 3.9). For windings in the axial direction, its equivalent thermal conductivity (4 W/mK) is severely impacted by insulations and the space occupied by air. Compared to the thermal conductivity of Cu (~379 W/mK), the difference is about 100 times. Even when water jacket cooling is applied between the stator and the housing, the overall thermal resistance is still high.

Along the axial side of the stator winding, thermal conductivity is much better (~389 W/mK), but dissipation capability at the end is limited since normally only natural air convection is available. In addition, the dissipation area in the axial direction is small. Heat transfer capability with natural air convection is very limited, with the HTC being ~10 W/m^{2}K. This is extremely low compared to the HTC of 4000 W/m^2*K with effective water jacket cooling. This is one major reason that

TABLE 3.3
Typical Thermal Properties in EV Traction Motors

	Thermal Cond. (W/mK)	Density (kg/m^3)	Spec. heat (J/kg*k)
Rotor Shaft	50	7978	480
Lamination	17–46	7600–7750	500
Windings	400	8978	385
Magnetics	10	7400	440–460
Insulation	0.1–3	900–1400	1500
Al housing	200	2700	921
Air Gap	0.026	1.009	1009
Coolant (Water)	0.47	1055	3062

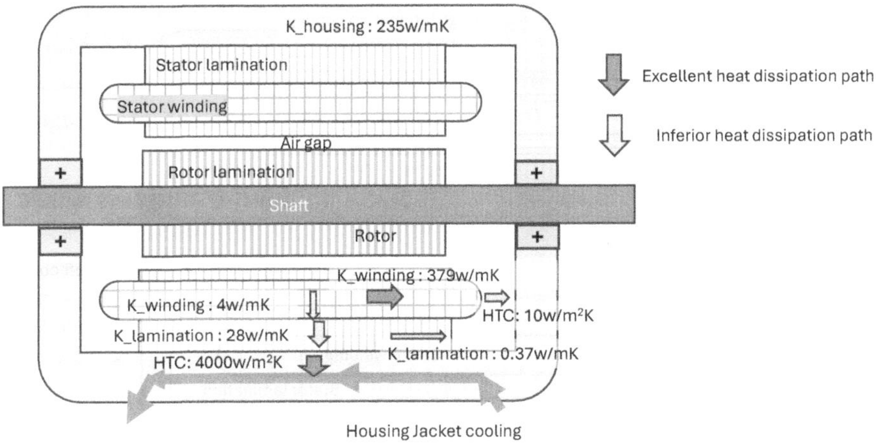

FIGURE 3.9 Illustration of heat dissipation characteristics of motor components along axial and radial directions based on Wang (2023).

spray cooling on the end winding is becoming a popular option to cool EV traction motors (Wang, 2023).

For the rotor in the radial direction, the heat dissipates through the air gap between the rotor and stator first, then to the stator, which is already hot. Air between the rotor ends and the housing makes rotor cooling even more difficult in the axial direction.

For stator and rotor laminations, it is difficult to dissipate heat along the axial direction. The thermal conductivity of lamination is low (~0.37 W/mK) along the axial direction compared to the value in the radial direction (~28 W/mK) since laminations are separated by insulations.

For traction motors, without active liquid cooling, the heat has to be dissipated mostly via air. Air cooling is not an effective cooling method to achieve high power density.

Liquid Cooling

Liquid cooling is widely adopted in traction motors to achieve higher power density (Gronwald, 2021). Liquid cooling of traction motors can be categorized into two types: direct cooling and indirect cooling. Indirect cooling is typically housing jacket cooling (shown in Figure 3.10). Cooling passages are designed within the motor housing close to the stator, through which cooling fluid passes to carry away the heat loss. Direct cooling involves injecting cooling fluid directly into the sealed motor, utilizing the high specific heat capacity of the cooling fluid for heat dissipation. In direct cooling, the coolant is in direct contact with hotspots such as the end stator windings, and the contact area can be significant. This results in much more efficient cooling. Another typical direct cooling method is shaft cooling (Bennion, 2015; Wang, 2021), where coolant is circulated inside the shaft. Through lamination cooling is also another type of direct cooling.

The effectiveness of the method depends on the nozzle used and the targeted region to be sprayed. Typically, the end of the winding is one target hot spot region to be sprayed on. Spray cooling on the end windings has significantly improved cooling efficiency because otherwise, the end windings have no short path to dissipate the heat to the low-temperature case except through the air (Nategh, 2014; Quassolo, 2022). In addition, this method directly cools the winding with minimal relevant thermal resistance. Oil spray cooling can achieve the HTC between 8000 and 10,000 W/m^{2}K (Ghahfarokhi et al., 2023). The effectiveness of oil spray is obvious. According to Toyota (Sano, 2016), by using oil spray cooling together with hairpin winding, the motor size was reduced by about 35%, with a 36% higher power density and 20% fewer losses.

However, introducing an additional spray system, including nozzles, increases the design complexity, and spray system optimization is needed to achieve optimum cooling performance. In some EVs, spray cooling is simplified and combined with other stator cooling methods. One example is the Lucid traction motor, where the spray is from the coolant exiting out of the stator lamination cooling channel (see Figure 3.10). This was used together with hairpin winding. Another example is Hyundai's E-GMP (Hyundai Motor Group, 2020) oil pipe design, where stainless steel pipes spray oil to cool the stator winding from the top. This simplifies the overall cooling system. In the Tesla Model 3, rotor spray cooling is used in addition to shaft cooling. The two ends of the shaft inside the motor have oil splashing holes, allowing oil to pass through shaft coolant channel and splash through the holes, cooling the rotor and stator inner windings directly (Lammotor, 2024).

Spray cooling is the most commonly reported method to cool the end of the stator winding. Spray cooling rotors will likely introduce more variances since the cooling efficiency is expected to vary with rotor speed.

Shaft Cooling

The air gap between the rotor and stator makes it difficult to dissipate the rotor heat loss in a radial direction. Shaft cooling is implemented to address this limitation. EV traction motors such as the Tesla Model S and Audi E-Tron use this method. There is a channel inside the shaft (see Figure 3.10) where the coolant flows. The heat from the rotor will be conducted toward the shaft first and then taken away by the coolant. This provides an effective liquid cooling method to cool the rotors. The design challenges are how to ensure the cooling channel is sealed and that the channel has minimal impact on the shaft's function and strength.

Other Cooling Methods

The temperature is not even in the stator. The variation can be very significant. Effective cooling design needs to consider this fact. As discussed earlier, one hotspot in the stator is end windings. The other one is the center of the stator slot, where low thermal conductivity materials are used to fill the stator slot (Nategh, 2014; Quassolo, 2022).

Lucid has developed one type of through lamination cooling method. This innovative microjet cooling system effectively removes heat from the stator winding, minimizing size and loss and boosting efficiency (Lucid, 2020, 2023). With this approach (Markus, 2021) (see Figure 3.10), the coolant is coming in from the middle (in the axial direction) of the stator and the housing. A microchannel is created in the stator lamination, through a narrow magnetic 'dead zone' between the windings along the axial direction. This allows the coolant oil to take away heat from the windings in very close proximity. Then the oil exits the channels and sprays onto the windings at the two ends. This innovation allows cooling of two hot spots with one cooling system. Combined with 900 V architecture, it enables the Lucid traction motor to achieve best-in-class power density.

Another example of through stator lamination cooling is the Tesla Model 3. Square holes are created in the stator core lamination along the outer radius, forming channels for the oil coolant to pass through along the axial direction (LamMotor, 2024). This avoids the need to design extra cooling channels in the housing and reduces the cost and size of the motor.

SUMMARY

Higher power density and torque density are always desired for traction motors to minimize mass and volume. A lighter motor can further increase efficiency and range. In general, several factors affect the power density of a motor (MPS, 2024):

1. Speed. Increasing motor speed can achieve higher power density without increasing torque or rotor volume.
2. Power Loss. Designing a more efficient motor with minimal loss from the beginning is the key to achieving optimum thermal performance for traction motors. This includes the adoption of various methods to reduce loss: selection of permanent magnetic motors, use of thin lamination, hairpin winding, and distributed winding.
3. Magnetic Materials. Motors using magnetic materials with higher magnetic flux density can have higher power density. This is the reason permanent magnetic motors use rare earth materials such as Neodymium and Samarium.
4. Winding. Higher winding densities will improve torque densities. Types of winding also matter. Hairpin winding is becoming more popular for EV applications. The winding method also affects power density.
5. Cooling. Power density also increases with better cooling (Kano, 2013). Liquid cooling is a natural choice over air cooling to achieve high power density. WEG is better than oil in terms of coolant performance. Oil cooling is gaining momentum in traction motor applications compared to WEG cooling because it can directly cool the hotspot.
5. DC Voltage. Power density (kW/kg) increases with higher DC bus voltage. This will be covered in Chapter 9 when 800 V is discussed.

BIBLIOGRAPHY

About-Motors. (2023). *Permanent magnet synchronous motor.* Retrieved from https://about-motors.com/motorcontrol/pmsm/

Akatsu, K. (2013). New trend of motor technology for automobiles - Introduction and overview. *2013 IEEE ECCE Asia Downunder.* IEEE.

Allegro MicroSystems. (2022). *EV traction motor requirements and speed sensor solutions.* Retrieved from https://www.allegromicro.com/en/insights-and-innovations/technical-documents/hall-effect-sensor-ic-publications/an296226-ev-traction-motor-requirements-and-speed-sensor-solutions

Anton Paar. (2024). *Viscosity of automatic transmission fluid - ATF.* Retrieved from https://wiki.anton-paar.com/us-en/automatic-transmission-fluid-atf/

Bennion, K. E. (2015, Jul 6). *Convective heat transfer coefficients of automatic transmission fluid jets with implications for electric machine thermal management.* Retrieved from Reprint from InterPack 2015: https://www.nrel.gov/docs/fy15osti/63969.pdf

Bourgault, A. E. (2019). A survey of different cooling methods for traction motor application. *2019 IEEE Canadian Conference of Electrical and Computer Engineering* (pp. 1–4). Edmonton, AB, Canada: IEEE.

Burress, T. E. (2013). Benchmarking EV and HEV power electronics and electric machines. *2013 IEEE Transportation Electrification Conference and Expo (ITEC).* Detroit, MI, USA: IEEE.

Donaldson, P. (2021). *Motor laminations.* Retrieved from https://www.emobility-engineering.com/motor-laminations/

Editorial Team. (2023, Feb 14). *What is an electric vehicle traction motor?* Retrieved from https://www.everythingpe.com/community/what-is-an-electric-vehicle-traction-motor#:~:text=Motor%20Control%20System%20of%20an%20EV&text=The%20motor%20controller%20regulates%20the,then%20supplied%20to%20the%20motor

Edmondson, J. E. (2022). *Electric motors for electric vehicles 2022–2032.* Retrieved from https://www.idtechex.com/en/research-report/electric-motors-for-electric-vehicles-2022-2032/842

EVANNEX. (2018, Sep 12). *Tesla's chief motor engineer discusses model 3 motor tech.* Retrieved from https://insideevs.com/news/339825/teslas-chief-motor-engineer-discusses-model-3-motor-tech/

Ghahfarokhi, P., et al. (2023). The oil spray cooling system of automotive traction motors: The state of the art. *IEEE Transactions on Transportation Electrification*, *9*(1), 428–451.

Gobbi, M. E. (2024). *Traction motors for electric vehicles: Maximization of mechanical efficiency – A review.* Retrieved from https://www.sciencedirect.com/science/article/pii/S0306261923018603

Goli, C. E. (2021). Review of recent trends in design of traction inverters for electric vehicle applications. *021 IEEE 12th International Symposium on Power Electronics for Distributed Generation Systems (PEDG)* (pp. 1–6). Chicago, IL, USA: IEEE.

Gronwald, P. E. (2021). Traction motor cooling systems: A literature review and comparative study. *IEEE Transactions on Transportation Electrification*, 7(4).

Hanejko, F. (2022, Nov 29). *Induction vs. permanent magnet motor efficiency | Electrification.* Retrieved from Horizon Technology https://www.horizontechnology.biz/blog/induction-vs-permanent-magnet-motor-efficiency-auto-electrification#:~:text=The%20inherent%20efficiency%20of%20a,set%20by%20the%20frequency%20drive.

Hughes, A. E. (2013). *Electric Motors and Drives: Fundamentals, Types and Applications* (4th ed.). Boston, MA: Newnes.

Husain, I. E. (2020). Electric drive technology trends, challenges, and opportunities for future electric vehicles. *Proceedings of the IEEE, 109*(6), 1039–1059.

Hyundai Motor Group. (2020, Dec 2). *Realizing new EV lifestyles: 5 advantages of the E–GMP.* Retrieved from https://www.hyundaimotorgroup.com/story/CONT0000000000089071

Jenkins, J. (2019, Mar 6). *A closer look at the losses in EV motors.* Retrieved from Charged ELECTRIC VEHICLES MAGAZINE https://chargedevs.com/features/a-closer-look-at-the-losses-in-ev-motors/

Jenkins, J. (2023, Sep 19). *How to improve EV traction motor efficiency.* Retrieved from Charged Electric Vehicles Magazine https://chargedevs.com/features/how-to-improve-ev-traction-motor-efficiency/

Kano, Y. E. (2013). Current specifications of vehicle motors. *2013 IEEE ECCE Asia Downunder.* Melbourne, VIC, Australia: IEEE.

Kato, T. E. (2013). Advanced Technologies for traction motor for Automobile. *2013 IEEE ECCE Asia Downunder.* Melbourne, VIC, Australia: IEEE.

Konovalov, D. E. (2023). Recent developments in cooling systems and cooling management for electric motors. *Energies 2023, 16*(19), 7006.

Krings, A. E. (2020). Review and trends in electric traction motors for battery electric and hybrid vehicles. *2020 International Conference on Electrical Machines (ICEM).* Gothenburg, Sweden: IEEE.

Laird Thermal. (2024). *Common coolant types and their uses in liquid cooling systems.* Retrieved from https://lairdthermal.com/thermal-technical-library/application-notes/common-coolant-types-and-their-uses-liquid-cooling-systems

LamMotor. (2024, Nov 9). *Comparison of motor stator and rotor technologies of Tesla, BYD, and Huawei.* Retrieved from https://lammotor.com/comparison-of-motor-of-tesla-byd-and-huawei/

Lammotor. (2024, Oct 24). *Tesla model 3 motors: Induction motor and permanent magnet synchronous motor.* Retrieved from https://lammotor.com/electric-motors-in-tesla/

Lucid. (2020, Sep 2). Retrieved from Lucid Motors' Proprietary Electric Drivetrain Technology Powers Record-Setting Performance and Industry-Leading Efficiency in the Lucid Air https://lucidmotors.com/media-room/lucid-motors-proprietary-electric-drivetrain-technology-powers-record-setting-performance

Lucid. (2023, Jan 25). *Lucid unveils state-of-the-art motorsports electric drive unit, taking Lucid air's advanced electric motor and drivetrain technology to the racetrack.* Retrieved from https://ir.lucidmotors.com/news-releases/news-release-details/lucid-unveils-state-art-motorsports-electric-drive-unit-taking

Marchio, C. (2024, Apr 16). *Permanent magnet motor in model 3 Tesla.* Retrieved from https://www.stanfordmagnets.com/permanent-magnet-motor-in-model-3-tesla.html

Markus, F. E. (2021, Nov 15). *2022 Lucid air: Take a deep dive into the stunning EV's powertrain tech.* Retrieved from MotorTrend https://www.motortrend.com/features/2022-lucid-air-powertrain-tech-deep-dive/

Mohan, N. E. (2002). *Power Electronics: Converters, Applications, and Design* (3rd ed.). Hoboken, NJ: Wiley.

MotorXP-PM. (2020, Jun). *Performance analysis of the Tesla model 3 electric motor using MotorXP-PM.* Retrieved from https://motorxp.com/wp-content/uploads/mxp_analysis_TeslaModel3.pdf

MPS. (2024). *General properties of electric motors.* Retrieved from https://www.monolithicpower.com/jp/learning/mpscholar/electric-motors/introduction-to-motors/general-properties-of-electric-motors?srsltid=AfmBOopyuSqYfPgnQQMP72qksjyVMt1gJdOoBZ6naHJRukxXrLmux4VE

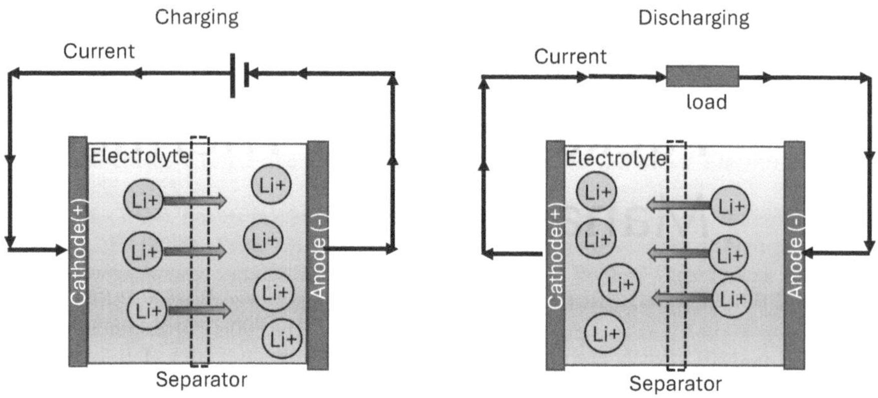

FIGURE 4.1 Illustration of lithium-ion battery charging and discharging.

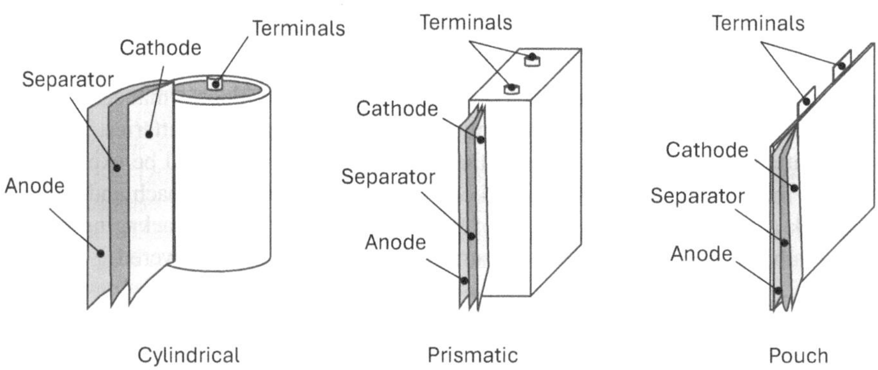

FIGURE 4.2 Three types of EV battery cell.

Charging and Discharging

When charging occurs (see Figure 4.1), an external voltage is applied and drives the lithium ions moving from the cathode through the electrolyte to the anode inside the cell. In the external circuit, the electrons move from the cathode to the anode, and the current is in the reverse direction. An electrochemical reaction happens during the process to convert electrical energy to chemical energy. In this way, the energy is stored in the battery.

Discharging is the reverse of charging (Figure 4.1). Now, chemical energy stored in the battery is converted to electrical energy and the battery supplies the power to drive the vehicles. During discharging, lithium ions move from the anode to the cathode through the electrolytes in the cell. In the external circuit, the electrons flow from the anode to the cathode and current flows from the cathode to the anode.

Battery Loss and Resistance

Batteries generate heat loss during charging and discharging. The typical efficiency of a battery system is around 85–95%. The heat loss is around 5–15%.

The heat loss originates from several sources.

1. Ohmic heat loss due to battery cell resistances and all the connections. The loss is proportional to the square of the current (I^2R) (I is the battery current). The internal battery cell resistance includes electrodes, electrolytes, and current collectors. The connections include terminal welding spots, wires, and connectors.
2. Electrochemical polarization
   Another source is polarization resistance (Rp), which accounts for ionic diffusion in the solid phase of the electrode materials and electrolyte.
3. Irreversible Reactions
   The loss is related to the charge transfer reaction occurring at the electrode/electrolyte interface at the beginning.

Two resistance parameters are used to represent the impact of those losses: direct current internal resistance (DCIR) and alternating current internal resistance (ACIR). DCIR is the resistance in charge and discharge when a DC is applied (Suffiyan, 2022). ACIR is the resistance in charge and discharge when an AC is applied. ACIR measurements are common and more standardized (Zollo, 2020). During an ACIR test, an AC of around 100 mA and 1000 Hz is applied; the real part of impedance (V_{ac}/I_{ac}) obtained is ACIR.

The factors affecting the internal resistance of a battery cell include current, state of charge (SoC), state of health, and temperature. Within the operating temperature range, the internal resistance of a cell drops as temperature rises. This is because the current generation in the battery is a chemical reaction. The reaction rate generally increases when temperature rises. Fully charged battery cells normally have lower resistance than discharged cells. As the battery ages, its internal resistance will increase due to material degradation and electrolyte breakdown. Additionally, battery resistance changes with loading conditions. For example, during normal cruising, low to medium discharge currents are used, the resistance is low and relatively constant over time. However, during acceleration or climbing, high battery current is generated and the resistance rises noticeably.

Battery Cell

The basic building block for current EV battery systems is a lithium-ion battery cell with a voltage around 3–4 V. There are three types of battery forms: cylindrical cell, pouch cell, and prismatic cell (see Figure 4.2).

Cylindrical cells look like AA batteries. They are one of the most popular forms of EV battery cells. Tesla, Rivian, Lucid, and other EV companies use this type of cell. In the cell, the anode, separator, and cathode are wound in a spiral and inserted into a metal cylindrical case with terminals at both ends. The cylindrical cells come in different dimensions. The 18650, 2170, and 4680 are the most popular (see Figure 4.3). The 18650 cell is 18 mm in diameter and 65 mm in length. It weighs 47 g and has a nominal voltage of 3.7 V. Each cell can have 1800–3500 mAh capacity. The 2170 cell has the same voltage as the 18650 cell but with a capacity of 4800 mAh. The 4680 cell's capacity is 26,000 mAh, which is much higher. The cell voltage varies with the SoC. A standard Tesla cylindrical cell has a voltage around 3–4.2 V. These cells are first connected in series and then in parallel to form a cell module with the required voltage and current rating.

Pouch cells directly pack both the electrode and separator into a rectangular-shaped pouch (see Figure 4.2). Without a metal case like the other two, pouch cells are flexible, light, and have the highest energy density. Of course, additional support is needed when pouch cells are integrated into the battery pack.

Prismatic cells are a group of pouch cells bundled together. Prismatic cells are normally contained in a rectangular metal case. They are much larger than cylindrical cells and have higher power density because less casing material is used. Some Audi models use prismatic cells.

Battery Performance Characteristics

A battery is typically characterized by its capacity, voltage, current, and power. Battery cell capacity defines the amount of energy a battery cell can store or deliver. It is measured in either ampere-hours (Ah) or milliampere-hours (mAh). Each type of cell has a specific amount of capacity. The 18650 cell's capacity ranges from 2500 to 3500 mAh. A cell with a 2500 mAh capacity means it can provide 2500 mA of current for 1 hour. Battery SoC is another important parameter. It is the percentage of energy remaining in a battery relative to its total capacity. SoC is a good indicator of battery health and performance.

The nominal voltage of a lithium-ion battery cell is typically around 3.6–3.7 V. One important characteristic of the lithium-ion battery is that its voltage correlates

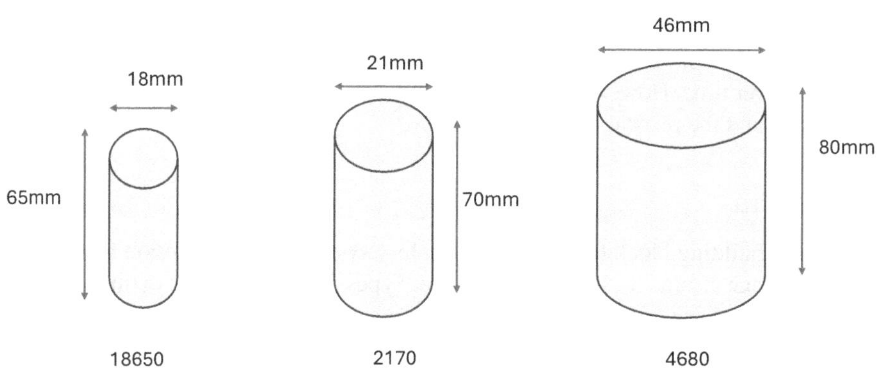

FIGURE 4.3 Dimensions of typical cylindrical cells.

to its SoC. For example, for a battery cell with 3.8 V nominal voltage, its voltage is 4.2 V at 100% SoC, and its voltage drops as SoC reduces. It drops to around 3.8 V at 50% SoC, and drops significantly as the SoC drops below 20%. Based on this characteristic, the BMS estimates SoC by measuring cell voltage.

The current of a battery cell is defined by the C-rate. The C-rate of a battery cell is defined as the ratio of charge or discharge current to its capacity. An EV battery is typically designed for 1C continuous discharge and 3C or higher discharge for a short period of time, such as 30 seconds. High discharge rates occur at rapid acceleration or while climbing steep slopes. For example, the current of a fully charged 2500 mAh battery cell rated at 1C means it can provide 2500 mA for 1 hour. A battery of the same capacity with a 3C rating means it can provide 7500 mA. One 18650 battery cell typically has a capacity between 2500 and 3500 mAh and a C-rate from 1C to 3C. If one 18650 cell with a 3500 mAh capacity and a voltage of 3.7 V is discharged at a 1C rate, the current is 3500 mA and the power output of the cell is about 13 W.

Maximum 30-second discharge current and maximum continuous discharge current are important parameters to characterize a battery cell's discharge capability. Maximum 30-second discharge current defines the maximum current that a battery can be discharged up to in 30 seconds. Along with the peak power of the traction motor, they define the acceleration performance of the vehicle (MIT Electric Vehicle Team, 2008). In comparison, maximum continuous discharge current defines the maximum current that the battery can be discharged at continuously. Along with the maximum continuous power of the traction motor and inverter, it defines the top sustainable speed of the vehicle (MIT Electric Vehicle Team, 2008). Maximum continuous discharge current of the battery system is limited by three factors (Nigel, 2023): cell chemistry such as cell ion charge transfer kinetics; thermal performance; and busbar continuous current rating. Hence, the thermal performance of the battery, including the maximum allowable cell temperature, determines the continuous power loss that a battery is allowed to produce without overheating. The maximum cell temperature gradient allowed and the maximum cell to cell temperature difference allowed are also important factors to be considered.

Generally, the maximum charging current for EV batteries is specified by battery cell manufacturers. Typical EV battery cells can charge/discharge at 1–3C current continuously. The maximum charging current for lithium-ion batteries is typically 1C or smaller under continuous conditions (X-Engineer, n.d.). Slower than 1C charging rate, such as 0.2C, is the C-rate recommended for use in actual applications. 0.2C means 5 hours to fully charge. Therefore, it is safer and preferred to do overnight charging. This helps to extend the battery's life (TYCORUN, 2023). The BMS typically controls the driving discharge rate between 0.2 and 0.5C (Etengoff, 2024).

The highest continuous charging rate occurs during DC fast charging. Most current EV DC fast charging is capable of charging from 10–20% SoC to 80% SoC in approximately between 15 minutes and 1 hour. This corresponds to a high C-rate (high charging current) of around 2–3C. For example, it takes 30 minutes to charge Tesla Model 3 (60 kWh) from 10 to 80% capacity. This corresponds to 1.4C. Fast charging (high C-rate) is always preferred by drivers. CATL introduced its new low-cost EV battery based on 4C ultra-fast charging. The battery can add 370 miles in 10 minutes (Johnson, 2024). While DC fast charging significantly reduces charging time to satisfy consumers' needs, it may potentially increase resistance and heat loss, sacrificing battery longevity and reliability.

Battery cell power is characterized by peak power and continuous power. Peak power is determined by the cell's maximum current and corresponding voltage. Peak power only lasts for seconds. Maximum continuous power is determined by the maximum continuous current and corresponding voltage.

Peak power and maximum continuous power are important parameters for battery packs. They describe how much power a battery pack can deliver under different driving conditions. Peak power describes the amount of power a battery pack can deliver in a short period of time (i.e., 10 seconds) (Nigel, 2023), which is normally the surge power. Peak power is critical for EV acceleration performance. The EV battery pack's peak discharging power is directly related to its individual battery cell's maximum discharging rate. Using the first-generation Tesla Roadster as an example, maximum peak output power for its battery is 210 kW at 411 V. This can only last 5 seconds (Tesla, 2024a). Maximum continuous power represents the sustainable power a battery pack can deliver for a long time continuously without overheating, sacrificing battery performance and reliability (DroneQuote, 2023). Maximum continuous power of the battery is critical to ensure an EV's capability to maintain top speed.

BATTERY MANAGEMENT SYSTEM

BMSs act as stewards of the battery system. The main function of the BMS is to monitor and control the state of the battery. It measures the voltage of individual battery cells, temperature, and pack current. It keeps all the cells in the battery pack under the same SoC throughout the battery's life (Panasonic, 2023).

A higher voltage of battery cells generally relates to a higher SoC. By measuring the voltage of each individual cell, the BMS determines the SoC for each cell. The BMS also senses each cell's temperature. It measures the current in and out of the whole battery pack and determines how much power and energy have been consumed or generated. Subsequently, as it monitors SoC, it discharges the cells with higher voltage and achieves a balanced SoC for all the cells. This protects battery cells from under-utilization, over-discharge, cell short circuits, and extreme temperatures.

A critical function of the BMS is to detect a battery system isolation fault and cut off the power supply when a problem occurs. An isolation fault occurs when the electrical isolation between EV HV components (such as battery cells, motors, and inverters) and the vehicle's chassis ground fails. The BMS continuously measures isolation resistance between the HV battery's positive and negative terminals and the chassis ground. Normally, the resistance is very high (>1 Mohm). If there is an isolation fault, the resistance will drop significantly. The BMS will detect it and open the contactors in the junction box. The BMS also monitors thermal management systems (such as the flow of coolant).

In summary, the BMS controls the battery junction box to prevent over-current, short circuits, and extreme thermal events, improving overall system safety.

BATTERY SYSTEM PACKAGING

The most common battery packaging solution is Cell → Cell Module → Battery Pack. A number of cells are first packaged into a cell module, then several cell modules are put together to form the battery pack (see Figure 4.4).

Most current EV battery packs have a 400 V nominal voltage, though there are some with 800 V. Based on the voltage and driving range needed, an appropriate combination of battery cells can be determined. Many cells are arranged in series/parallel to provide sufficient power and energy to EVs.

Individual cells are first connected either in series or in parallel, to form a cell module with specific module voltage and energy capacity. Each cell module is enclosed in a metal housing to protect the cells from vibration and other environmental impacts. Next, several cell modules are assembled to form the battery pack (see Figure 4.4). The battery pack provides the desired voltage, current, and power output.

The example below is the battery cell module used in one EV. It is a 62 kWh version, where 12 prismatic cells with a voltage of 3.7 V are connected in series, and then two sets are connected in parallel. Therefore, the cell module voltage is

$$3.7V \times 12 = 44.4V$$

In total, there are 24 prismatic cells packaged in one cell module. There are nine cell modules connected in series in a battery pack. Therefore, the battery pack voltage can be determined.

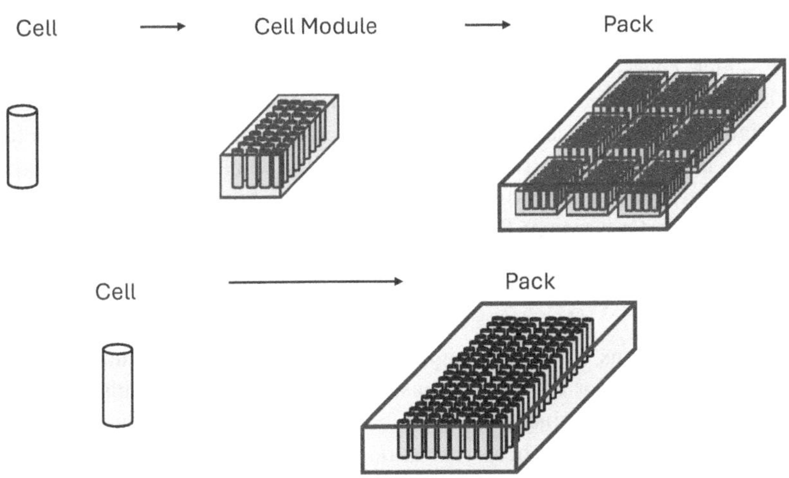

FIGURE 4.4 Illustration of battery cell to pack assembly with cylindrical cells.

$$44.4V \times 9 = 399.6V$$

Tesla vehicles' battery packs are a cell-to-pack design (shown in Figure 4.4). They do not have a dedicated cell module housing. All the cells are placed next to each other. This approach eliminates additional weight (spaces) and assembly steps associated with cell modules, resulting in higher energy density and lower cost for the entire battery pack system. One additional advantage of this approach is that the battery pack can also provide structural support to the whole vehicle (Donaldson, 2021).

A case study is used here to explain Tesla battery pack structure and configuration based on the Tesla Model S (Bhowmick, 2021), where a 18650 cell is the basic building block. Each 18650 cell has a nominal voltage of ~3.8 V, a capacity of 3.4 Ah, and an energy of 12.4 Wh. Its continuous current rating is 4.89 A. The battery pack includes 16 cell modules in series without dedicated module housing. Each cell module has 6 cells in series and 74 cells in parallel. Therefore, the battery pack configuration is 6×16 cells in series and 74 cells in parallel.

$$\text{Pack voltage is } 3.8\text{V} \times 6 \times 16 = 364.8\text{V}$$

$$\text{Pack current is } 4.89\text{A} \times 74 = 361.9\text{A}$$

$$\text{Pack capacity is } 12.4Wh \times 6 \times 16 \times 74 = 88kWh$$

For EVs, the battery system occupies a significant portion of weight and volume. The Tesla Model 3's battery weight is about 1000 lb, while the whole vehicle weight is around 4000 lb. About a quarter of EV weight goes to the battery. Using this battery to provide structural stiffness for the whole vehicle makes much better use of the battery system (Bower, 2022). Tesla's 'structural battery' integrates the battery into the vehicle's chassis. Vehicles typically need certain torsional rigidity, which is defined as the torque needed to produce a unit angle of twist. Higher torsional rigidity means responsive steering, a good driving experience, and better ride quality. The traditional vehicle chassis is the major component determining vehicle torsional rigidity. In a traditional vehicle, there is a structural floor pan and rockers to provide structural rigidity. In Tesla's design, the battery pack is part of the vehicle structure. It provides a large portion of structural stiffness for the whole vehicle. This is expected to significantly reduce the additional mass and space needed to design a dedicated chassis structure for rigidity (The Limiting Factor, 2021).

THERMAL RUNAWAY

Safety considerations to contain high thermal instability of lithium-ion batteries are always a priority for EV applications. Thermal runaway in lithium-ion batteries occurs when temperature increases in the battery, resulting in further reactions. This reaction leads to even more heat generation than what the cooling system can handle. Therefore, the temperature increases even more, eventually resulting in a catastrophic failure. Thermal runaway can result in cell venting, fire, or even explosions.

As Warner (Warner, 2019) explains, once the temperature of the cell reaches ~80°C, the solid electrolyte interface (SEI) protective layer on the anode will start to break down and result in a reaction of lithium with the electrolyte solvent, generating additional heat. This further increases temperature. As the temperature reaches 100–120°C, the electrolyte starts to break down and gas is produced. As the temperature continues to increase to around 120–130°C, the separator layer starts to melt. Next, the cathode layer starts to break down as the temperature further increases to around 130–150°C. This comes with oxygen generation. All of these are exothermic reactions with significant heat generation, eventually driving the cell to failure and fire. This is why it is desired to have the battery temperature well below 80°C, preferably below 60°C. For some EV batteries, the recommended temperature is 0–45°C for charging, and –20–60°C for discharging (Prismatic & Pouch Battery Packs, n.d.).

Root causes for thermal runaway include: overcharging, over-discharging, physical damage to the battery, high temperature or hotspot, and manufacturing defects. Overcharging and over-discharging are two main causes. Therefore, SoC is a critical cell parameter. The SoC status of each battery cell is related to its voltage level. For example, a fully charged lithium-ion 18650 cell has a voltage around 4.2 V. If charging continues after the battery is fully charged, more energy will flow to the cell. The battery voltage can continue to increase, and the cell enters the 'overcharging stage,' where excess energy is stored in the cell. This will cause heat generation and eventually thermal runaway. Over-discharge happens when discharging continues even after the battery cell voltage drops below cut off voltage (~3.0 V/cell for typical lithium-ion batteries). Over-discharging can cause copper in the anode to dissolve, creating a short circuit between the anode and cathode when the battery is charged again. This could lead to battery cell damage and thermal runaway. Therefore, a new EV battery is recommended to be charged up to 80% capacity and to stop discharging before reaching 30% capacity to reduce battery stress and the possibility of thermal runaway. Keeping the battery level above 30% and only charging it to 80% is essential to maintain the pack's longevity. Physical damage to a battery can cause internal short circuits and induce thermal runaway as well.

Proper thermal management and protective circuitry are essential to prevent thermal runaway. The first solution is to avoid overcharging and over-discharging by the BMS. One major function of the BMS is to evaluate the SoC for each battery cell by monitoring its voltage and temperature in real time. Through this data, the BMS can ensure the normal operation of healthy cells, detect abnormal cells, and regulate or even limit the operation of suspected cells. The second solution is to design appropriate thermal management to avoid high temperatures and high temperature gradients for battery systems. Most EV battery systems use active liquid cooling systems. With the help of the BMS, the cooling system will try to maintain a desired temperature within a battery cell and minimize temperature difference from cell to cell. Another solution is isolation, a passive but least expensive method. The goal is to contain thermal runaway within each cell, long enough to ensure thermal runaway won't propagate from one cell to nearby cells too quickly. In the end, passengers will have enough time to escape. One common escape-time requirement is 5 minutes. Thermal runaway for a single cell lasts from 20 seconds to a minute. Without any method to delay propagation, it won't take long before thermal runaway propagates and ultimately the whole battery system fails (Koch, 2018). One isolation option is to have a battery pack enclosure that is heat and fire resistant. Another option is to use some type of insulation between cells that makes it more difficult to transfer heat from the hot cell to nearby cells.

BATTERY SYSTEM THERMAL MANAGEMENT

EV batteries need to last 10 years/100,000 miles despite their degradation over time. A good thermal management system and BMS can extend battery life while ensuring battery safety, performance, and efficiency.

Thermal Management Objective

The primary objectives of battery thermal management include (Warner, 2015; Massobrio, 2023):

1. Keep the battery temperature within an optimal range.
   Lithium-ion chemistries tend to have optimum performance at temperature ranges from 10 to 35°C (Wang, 2016). Most lithium-ion batteries' temperature operating range is –20–45°C. Between –20 and –40°C, the electrolytes may start to freeze, and the battery performance will be limited. Above 60°C, as previously discussed, the cell chemistry will start to deteriorate. The Tesla Model 3 Owner's manual (Tesla, 2024b) recommended 'For better long-term performance (for the battery), avoid exposing Model 3 to ambient temperatures above 140°F (60°C) or below –22°F (–30°C) for more than 24 hours at a time.' Audi (2021) stated 'The battery's feel-good temperature is between 30 and 35°C, and its operating range extends from minus 30 to plus 50°C.'
2. Ensure uniform temperature distribution across the battery pack.
   One recommendation is to ensure the temperature gradient from module to module is less than 5°C (Wang, 2016). Temperature variation across the cell is less than 3°C and from cell to cell is less than 2–3°C. This is because local temperature hotspot likely results in accelerated chemical reactions at the cells and leads to thermal runaway.
3. Eliminate temperature-related safety issues such as thermal runaway.

To summarize, the thermal challenges of the battery system are to maintain it within a very narrow range of 10–35°C for most of its use life and to ensure minimal temperature gradient from cell to cell and within the cell. This is challenging since sometimes the ambient temperature alone already exceeds this range. In addition, the heat loss generated during battery charging and discharging needs to be dissipated effectively to minimize temperature rise. For a 100 kW battery system, heat loss is about 10 kW, assuming 10% heat loss. If not properly managed, the heat loss can increase battery temperature, degrade its performance, reduce its lifespan, and even introduce safety issues. It has been reported that without an active cooling system, the 2012 Nissan Leaf battery system experienced a loss of battery capacity after only a short period of time when vehicles drove under hot climates for about 150% of the average US annual mileage (Warner, 2015).

In the current EV battery system, active cooling with water/glycol is the default choice. The selection of the cooling system depends on the usage cycles, battery cell form, and ambient conditions.

Heating

Active heating is necessary for the battery system when outside temperatures drop below minimum operational temperature (8°C). As temperature drop around 8°C or lower, liquid electrolyte viscosity increases, leading to slower ion movement and lower power delivery. The ID4 battery system is actively heated by the Z132 heating element when the battery system temperature drops below 8°C (Volkswagen Academy, 2020). A coolant heater is used in the Tesla Roadster. The coolant heater is only activated to raise the internal temperature of the battery above 8°C before charging commences. The heater will only activate if the charge port door is open (Tesla, n.d.). In Tesla Model 3, there is no dedicated heater. The waste heat from the motor and inverter is used to heat the battery in cold weather (Bower, 2018). BMW has a resistive heater under the battery pack to heat the battery system.

Bottom Cooling

Bottom cooling is the typical cooling option for the cell to module to battery pack packaging. The bottom of the battery pack is a large cooling plate, to which the bottom of the cell modules is thermally attached via some type of thermal interface material (TIM). This approach is used in the Porsche Taycan, VW IDx, BMW i4, and Audi E-Tron (Munro Live, 2024) .

VW ID.4 uses battery cells made by LG Chem. The cells are in pouch form. They are first thermally connected to the bottom of the battery module with a TIM; then, the module base is thermally connected to the battery pack cooling plate underneath the pack (see Figure 4.5) (Nigel, 2022).

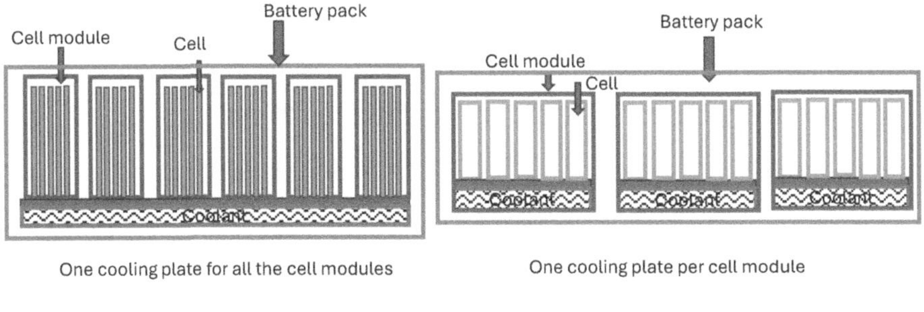

FIGURE 4.5 Two types of bottom cooling approaches for battery pack systems with cell modules.

Similarly, an indirect cooling system is used under the Audi battery module (Audi, 2021). There is 'a compound structure of flat extruded sections through which a glycol/water mixture flows that circulates in its own low-temperature circuit. The temperature is transferred between the cooling plates and the cell space via a heat-conducting paste.'

Lucid Air uses one cooling plate for each cell module instead of one cooling plate for all cell modules. The cooling plate is integrated with each cell module from the bottom (see Figure 4.6 and Figure 4.7)

Side Cooling

This method is used in the Tesla battery system with cylindrical cells (Munro Live, 2024). Shown in Figure 4.7, the cold plate is sandwiched between two rows of cells. Coolant flows inside the cold plates adjacent to the sides of the cells. The cold plates are curved around the perimeter of the walls. Thermal paste is applied to fill the gap between the cold plates and the cells.

FIGURE 4.6 Lucid cell module where the cooling plate is located at the bottom (photo by Author).

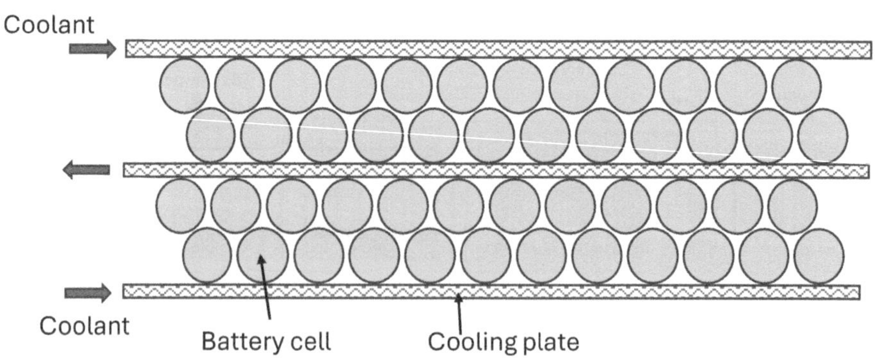

FIGURE 4.7 Illustration of side cooling.

Comparison between Two Battery System Cooling Methods

The choice between side cooling or bottom cooling depends on the battery system packaging. If cell modules are used, as in most EVs, bottom cooling is the choice. Tesla uses side cooling because its cells are directly integrated into the battery pack.

The effectiveness of these two cooling options is compared here. Regardless of cylindrical cell or pouch cell, the cell is made of one layer of cathode electrode and one layer of anode electrode, separated by the electrolyte and the separator. As electrodes are made from sheets of aluminum and copper, both have excellent thermal conductivity (100–400 W/mK). In contrast, both the electrolyte and separators are poor thermal conductors. Therefore, heat conduction along the in-plane layer direction is much higher than in the out of plane direction, where the heat has to dissipate through layers of low thermal conductivity layers. As Yu (2023) has shown, thermal conductivity of the 18650 cell is 14.2 W/mK in axial direction vs. 1.045 W/mK in radial direction, which is around a 14 times difference. A similar ratio appears for 2170 cells. However, due to the short length (L) and much bigger contact area (A) in the radial direction, the overall thermal resistance ($R_{th}=L/(K*A)$) for the battery cell in the radial direction (side cooling) can be comparable to the axial direction (bottom direction) or even better (Bower, 2019; Nigel, 2022).

Lastly, battery system cooling is different from other powertrain component cooling. It has lower power loss density and does not necessarily need high cooling efficiency in terms of thermal resistance. For example, for a 3500 mAh, 0.5C 18650 battery cell with a 3.7 V normal voltage and a 1.75 A current, the power of the cell is ~13 W. Assuming 95% efficiency, heat loss is around 0.65 W. The loss (0.65 W) for this dimension (diameter: 18 mm, length: 65 mm) is negligible compared to loss density of traction inverter devices (100 W loss over 10 mm×10 mm×0.07 mm). The loss density ratio is about 1:3.6E5 (4E–5 w/mm^3 vs. 14w/mm^3). As a result, the requirement to lower the thermal resistance is less demanding for battery system cooling. Instead, the objective for battery system cooling is to maintain all the cells at the optimum temperature range with a very small temperature gradient.

TECHNOLOGY TREND

Despite the fact EVs with lithium-ion batteries are gaining significant market share, the cost, charging speed, and safety of lithium-ion batteries are still major concerns. The key to address those concerns lies in battery cells with higher energy density, faster charging speeds, and greater safety. Tesla has made several important technological advancements. One of them is the tabless lithium-ion battery cell. As described in the Tesla patent (US Patent No. US20200144676A1, 2019; The Limiting Factor, 2020; Arar, 2020) for lithium ion prismatic cells, instead of a single small tab (see Figure 4.8), multiple tabs are used as the current collector. This significantly reduces the path from terminals to active electrode materials compared to the traditional single tab. This approach reduces cell resistance and improves thermal performances. As Christensen (2022) stated, this potentially cuts the total electrical

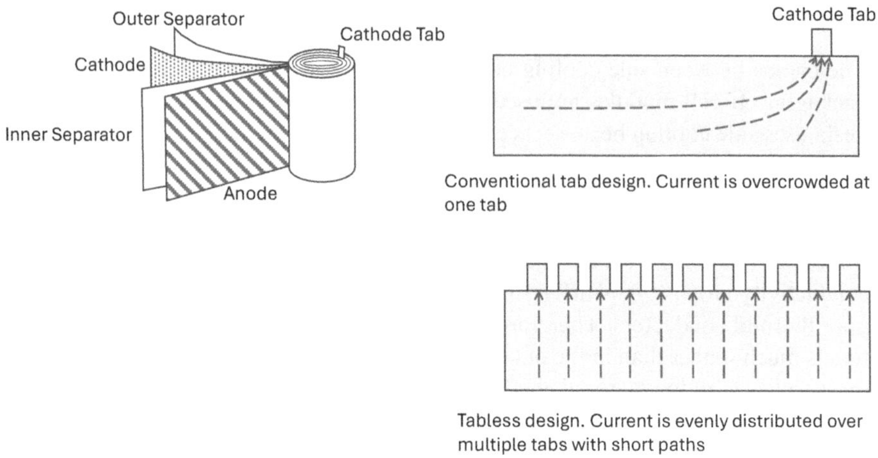

FIGURE 4.8 Illustration of tabless battery cell design.

path length from ~3.5 m to ~75 mm (almost 50 times shorter) in a 4680 cell, resulting in lower cell internal electrical resistance and reduced heat loss. He predicted a 10% increase in power density in 18650 cells with this design. Tesla stated this significantly increases its battery cell power-to-weight ratio and power rating (The Kilowatts, 2021). Tesla's 4680 cell with a tabless design offers 6× power and 5× energy capacity compared to its previous cells (Lambert, 2020).

Another trend is solid state batteries, which have the potential to significantly increase energy density. The major difference is that solid state batteries use solid conductors instead of liquid or gel-like electrolytes.

BIBLIOGRAPHY

Arar, S. (2020, Oct 7). *Just How Much of a "Breakthrough" Is Tesla's Tabless Battery Cell?* Retrieved from https://www.allaboutcircuits.com/news/just-how-much-breakthrough-teslas-tabless-battery-cell

Audi. (2021, Mar 3). *Battery and Thermal Management.* Retrieved from https://www.audi-mediacenter.com/en/emotive-design-and-revolutionary-technologythe-audi-e-tron-gt-quattro-and-the-audi-rs-e-tron-gt-13655/battery-and-thermal-management-13784#media-contacts

Bhowmick, S. (2021, Dec 16). *Tesla Model S Battery System: An Engineer's Perspective.* Retrieved from https://circuitdigest.com/article/tesla-model-s-battery-system-an-engineers-perspective

Bower, G. (2018, Dec 14). *Tesla VS. GM: Battle Of The Battery & Electronics Cooling Systems.* Retrieved from https://insideevs.com/news/341575/tesla-vs-gm-battle-of-the-battery-electronics-cooling-systems/

Bower, G. (2019, Oct 07). *Tesla Model S P100D+ Plaid: Everything We Know.* Retrieved from https://insideevs.com/photos/683100/insideevs-us-single-photo-upload/#4385180_tesla-model-s-plaid-analyzed

Bower, G. (2022, Jan. 25). *Tesla Model Y Structural Battery Pack: What We Know (And Don't Know).* Retrieved from https://insideevs.com/news/563013/tesla-structural-battery-pack/

Christensen, J. (2022, May 20th). *Tesla Tab Talk.* Retrieved from BOSCH Research Blog: https://www.bosch.com/stories/tesla-tab-talk/

Donaldson, P. (2021). *Cell-to-pack batteries.* Retrieved from https://www.emobility-engineering.com/cell-to-pack-batteries/

DroneQuote. (2023). *Battery 101: Why Peak Power and Continuous Power Output Matter.* Retrieved from Youtube: https://www.youtube.com/watch?v=uSMicefnjfw

Etengoff, A. (2024, Feb. 6). *How Do EV Battery Cell Charge and Discharge Rates Compare?* Retrieved from https://www.evengineeringonline.com/how-do-ev-battery-cell-charge-and-discharge-rates-compare/

Johnson, P. (2024, Apr 25). *CATL Unveils World's First LFP Battery with 4C Ultra-Fast Charging for 370-mi in 10 mins.* Retrieved from https://electrek.co/2024/04/25/catl-unveils-worlds-first-lfp-battery-4c-ultra-fast-charging/

Koch, S. (2018). Fast Thermal Runaway Detection for Lithium-Ion Cells in Large Scale Traction Batteries. *Batteries, 4*(2), 16.

Lambert, F. (2020, Sep 22). *Tesla Unveils New 4680 Battery Cell: Bigger, 6x Power, and 5x Energy.* Retrieved from https://electrek.co/2020/09/22/tesla-4680-battery-cell-bigger-power-energy/

Massobrio, A. (2023). *The Impact of Efficient EV Battery Thermal Management.* Retrieved from https://www.neuralconcept.com/post/the-impact-of-efficient-ev-battery-thermal-management

MIT Electric Vehicle Team. (2008, Dec.). *A Guide to Understanding Battery Specifications.* Retrieved from https://web.mit.edu/evt/summary_battery_specifications.pdf

Munro Live. (2024). *Keep Your Cool: Electric Vehicle Battery Cooling Methods.* Retrieved from https://www.youtube.com/watch?v=9DNe4gi7NYs

Nigel. (2022, Jun 28). *Thermal Conduction in a Cell.* Retrieved from Battery Design from Chemistry to Pack: https://www.batterydesign.net/thermal-conduction-in-a-cell/

Nigel. (2023, May 15). *Peak vs Continuous Power.* Retrieved from Battery Design from Chemistry to Pack: https://www.batterydesign.net/peak-vs-continuous-power/

Panasonic. (2023, Oct 20). *What Is BMS in an Electric Vehicle (EV)? - System That Monitors and Controls Battery State and Performance.* Retrieved from Panasonic: https://industrial.panasonic.com/ww/ss/technical/ap8

Prismatic & Pouch Battery Packs. (n.d.). Retrieved from EPEC Engineered Technologies: https://www.epectec.com/batteries/prismatic-pouch-packs.html

Suffiyan, M. (2022, Oct 29). *Internal Resistance: DCIR and ACIR.* Retrieved from https://www.batterydesign.net/dcir-acir/

Tesla. (2024a). *Power Electronics Module.* Retrieved from Roadster D&O: https://service.tesla.com/docs/Public/Roadster/TheoryOp/1.2.5/tabmenuframe.php-ID=DO.html

Tesla. (2024b). *Tesla Model 3 Owner's Manual.* Retrieved from https://www.tesla.com/ownersmanual/model3/en_us/GUID-7FE78D73-0A17-47C4-B21B-54F641FFAEF4.html

Tesla. (n.d.). *Tesla Roadster D&O.* Retrieved from https://service.tesla.com/docs/Public/Roadster/TheoryOp/1.2.5/tabmenuframe.php-ID=DO.html

The Kilowatts. (2021, Sep 23). *Tesla Battery Day - 20 Minute SUPERCUT.* Retrieved from https://www.youtube.com/watch?v=H8Vv-Dv6-k8

The Limiting Factor. (2020). *Tabless Electrode Patent (Deep Dive).* Retrieved from https://www.youtube.com/watch?v=Rcw3vA0gAMs

The Limiting Factor. (2021, Apr. 28). *#15 The Engineering of Tesla's Structural Battery Pack.* Retrieved from https://www.youtube.com/watch?v=f1Q3ccjhvqQ

Tsuruta, K. E. (2019). *US Patent No. US20200144676A1.* Retrieved from https://patents.google.com/patent/US20200144676A1/en

TYCORUN. (2023, Aug 2). *Maximum Charging Current - The Key Factor to Improve Battery Charging Efficiency.* Retrieved from https://www.tycorun.com/blogs/news/maximum-charging-current#:~:text=The%20maximum%20battery%20charging%20current,and%20a%20maximum%20of%20400mA1

Volkswagen Academy. (2020). *The High-Voltage System in the ID.4.* Retrieved from https://static.nhtsa.gov/odi/tsbs/2021/MC-10186407-0001.pdf

Wang, Q. E. (2016, Oct). *A Critical Review of Thermal Management Models and Solutions of Lithium-ion Batteriesfor the Development of Pure Electric Vehicle.* Retrieved from Renewable and Sustainable Energy Reviews: https://nottingham-repository.worktribe.com/output/974909/a-critical-review-of-thermal-management-models-and-solutions-of-lithium-ion-batteries-for-the-development-of-pure-electric-vehicles

Warner, J. T. (2015). *The Handbook of Lithium Ion Battery Pack Design.* Retrieved from https://pspowers.com/wp-content/uploads/2021/05/Handbook-of-lithium-ion-battery-pack-design-_-chemistry-components-types-and-terminology-PDFDrive-.pdf

Warner, J. T. (2019). *Lithium-Ion Battery Chemistries, Chapter 3.* Retrieved from https://www.sciencedirect.com/topics/chemistry/thermal-runaway

X-Engineer. (n.d.). *EV Design – Battery Calculation.* Retrieved from https://x-engineer.org/ev-design-battery-calculation/

Yu, H. E. (2023). Thermal Parameters of Cylindrical Power Batteries: Quasi-Steady State Heat Guarding Measurement and Thermal Management Strategies. *Applied Thermal Engineering, 231*(Aug).

Zollo, B. (2020, Apr 15). *Measuring DCIR of Lithium-Ion Cells.* Retrieved from https://www.electronicdesign.com/technologies/test-measurement/article/21128843/keysight-technologies-measuring-dcir-of-lithium-ion-cells

5 Thermal Design of Power Electronic Components

ABSTRACT

Power electronic components are critical electric drive system components that control the energy flow within the vehicle. They include traction inverters, OBCs, and DC-DC converters. One major EV power electronic components development trend is to achieve high power density. Power density is dictated by the amount of power applied to the component, the component's efficiency, and thermal performance. Power and efficiency determine the amount of power loss. Thermal performance determines how much power loss the components can handle without overheating or affecting component reliability. This chapter discusses the system design of inverters, OBCs, and DC-DC converters, where thermal design will be emphasized. State-of-the-art EV power electronic packaging technologies including double side cooling will be revealed.

INTRODUCTION

Power electronic components manage and regulate electrical energy flow within the vehicle. The major limiting factor in achieving high power density is that the heat generated in these components cannot be dissipated effectively because the component volume (area) is too small and/or the power loss is too high. Power semiconductor devices are the major source of power loss for EV power conversion systems. They are packaged either in discrete packages or in multichip power modules. A traction inverter power module typically needs to handle 100–200 kW of power (Goli, 2021). Even if inverter efficiency is 98%, the power loss (2–4 kW) is significant. One key target for power electronic components' packaging design is to achieve low thermal resistance, ensuring the junction temperature of these devices is within the limit, with minimal component volume/weight.

POWER ELECTRONIC COMPONENTS

Inverters, OBCs, and DC-DC converters are important EV power electronic components for managing power conversion and distribution in the vehicle.

INVERTER

The typical topology of traction inverters is shown in Figure 5.1. It includes a three-phase power module packaged with six sets of power devices, a gate drive board, and

DOI: 10.1201/9781003569961-5

Thus, inverter design needs to consider several worst-case scenarios, both thermally and electrically. These scenarios include maximum AC output (torque), maximum speed, maximum DC from the battery, and maximum loss in critical weak points such as power switches.

Reliability, size, weight, and cost need to be considered in inverter design as well (Blaabjerg, 2021; Mainka, 2011).

OBC

Typical bidirectional OBC topology includes an input EMI filter, a power factor correction (PFC) AC-DC converter, and DC-DC converters. The DC-DC converter includes a transformer, primary side, and secondary side (McDonald, 2022) (see Figure 5.2).

A PFC AC-DC converter converts AC from the electric grid to DC using rectification and PFC. PFC rectifies the input voltage to ensure the power factor is as close to unity as possible. (Here, the power factor is defined as the ratio of true power (*W*) to apparent power (*VA*).) This is achieved by incorporating elements such as inductors. Depending on the phase of the input power, it can be a one-phase or three-phase AC/DC converter.

The DC-DC converter adjusts the produced DC voltage before it goes to the battery. A high frequency (HF) transformer is used between the primary side and secondary side to provide HV isolation. The two most popular DC-DC converter topologies are the Inductor-Inductor-Capacitor (LLC) converter and the dual active bridge (DAB) converter (ROHM CO., LTD, 1997–2024).

OBCs can control the current and voltage output so the battery can be charged with either constant current or constant voltage to achieve optimum charging performance.

In PFC, the primary side and secondary side of DC-DC, power switches and diodes are used extensively. MOSFETs are the dominant choice for bidirectional OBC converters due to their wide availability and power capability (Yuan, 2021). WBG devices are becoming more attractive for OBC applications since they can be used under high switching frequencies. A higher switching frequency for OBCs

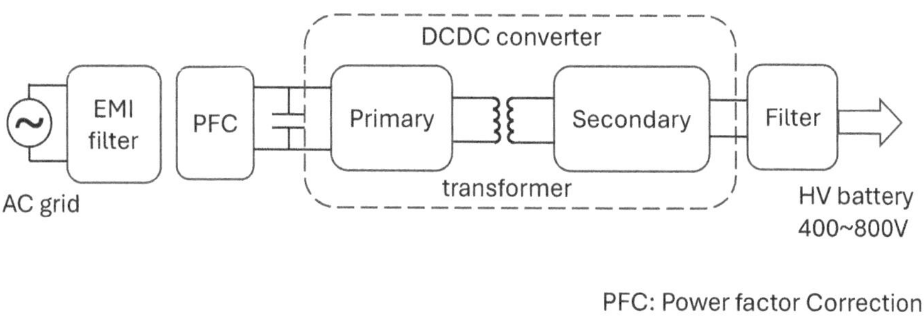

FIGURE 5.2 Typical simplified bidirectional OBC block diagram.

allows for smaller magnetic components, which enables higher power density. Discrete power electronic packages instead of power modules are the main choice for OBCs because of cost advantages and wide availability of discrete power devices.

Liquid cooling with a customized Al heat sink is the typical cooling choice for today's EV OBC power devices. One popular option is TO220 type packages attached to the heat sink with a thermal pad. Another option is surface mount packages such as D2PAK mounted on insulated metal substrate (IMSs), then coupled to the heat sink with a thermal pad.

The transformer provides galvanic isolation between the AC input (grid) and the DC output (EV battery system). It is a critical element for both LLC and DAB topologies. Customized transformer design is necessary to achieve an optimum design, which needs to consider safety, efficiency, and size reduction.

EMI is always a concern associated with EV OBC. The EMI input filter is one effective way to reduce EMI before it enters the grid or propagates through the EV electrical system. It filters unwanted noise from AC input power and protects the OBC by limiting the voltage and current spikes (McDonald, 2022). It employs both CM and DM filters. After DC-DC conversion, DC current with an adjusted voltage level is obtained. However, it still has many harmonics. Output filters are placed to reduce the remaining harmonics before the power goes to the battery.

Similar to inverters, EV OBCs require high power density and maximum efficiency. With higher efficiency, a smaller OBC is possible due to less power loss. Both LLC and DAB topologies in DC-DC have been adopted due to their capabilities in achieving high efficiency. Certainly, the efficiency varies with load conditions such as charging and discharging speeds. Passive components are heavy and have a low power factor. The use of passive components in PFC is generally inefficient but unavoidable for PFC functions. It generates a significant amount of power loss. Therefore, OBC efficiency is normally less than 95%. Although, for OBCs with SiC, >96% efficiency has been reported (Wolfspeed, 2022).

DC-DC Converter

The DC-DC converter converts HV DC from the battery to LV DC such as 12 or 48 V (see Figure 5.3) (Panasonic, 2023; Skyworks, 2024; Infineon, 2024). The HV from the battery first goes through the input filter to reduce EMI noise. Next, the DC link capacitor is used to reduce the voltage ripple. A transformer is used to ensure galvanic isolation between HV and LV. The output EMI filter is placed at the end. Another DC link capacitor is used to smooth voltage output. EMI filters are critical

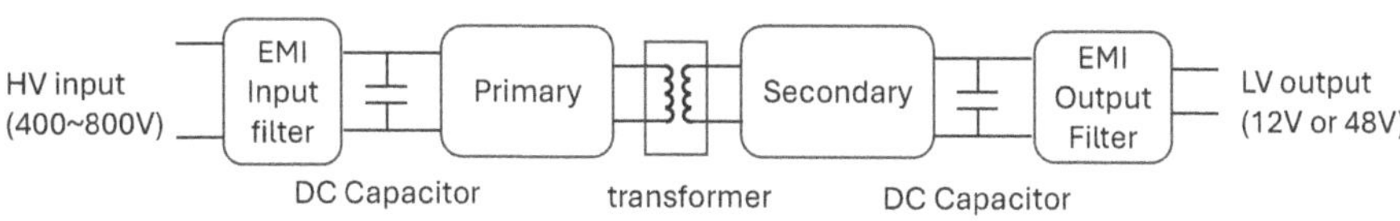

FIGURE 5.3 Block Diagram of HV-LV DC-DC converter.

for DC-DC converters to suppress internal and external EMI noise to meet EMI requirements.

DC-DC converters typically have power ratings around 2–3 kW and efficiency around 92–95%. Switching frequencies can be as high as 100 kHz to minimize transformer size. For most EV DC-DC converters, both primary side and secondary side switches use discrete packages and require liquid cooling (Donaldson, 2021). Like other EV power electronic components, DC-DC converters need to be compact, lightweight (higher power density), reliable, efficient, and cost effective.

EFFICIENCY AND POWER LOSS

Achieving high efficiency is a key target for inverters, OBCs, and DC-DC design. Higher efficiency leads to higher current/power capability or a more compact cooling design. 1% higher in overall motor and power electronic (mainly inverter) efficiency will save 1% battery capacity (Yildirim, 2014). Higher efficiencies indicate lower power loss, which allows for a smaller and lighter cooling system and higher reliability.

Most of the power loss in the inverter, OBC, and DC-DC converter occurs in the power switches and diodes. The power loss of power semiconductor switches typically includes conduction losses, switching losses, and other parasitic effects.

Conduction loss (P_{cond}) occurs when the power semiconductor switch is on and the device is in full conduction. The loss is caused by the on-resistance of the devices. As the amount of power transferred increases, the conduction current will increase so as the loss.

$$P_{cond} = I_D^2 R_{on} \tag{5.1}$$

Here,

I_D is current flowing through the switch. For MOSFETs, it is drain to source current.

R_{on} is on-resistance of the switch. The value is temperature dependent (see Figure 5.4).

Switching loss occurs during the transition between on and off. The switching loss of MOSFETs can be estimated (Keim, 2024; ROHM Semiconductor, 2016):

$$P_{switch} = \frac{1}{2} V_{in} I_D \left(t_R + t_F \right) F_{sw} \tag{5.2}$$

V_{in}: input voltage used to drive switched current through MOSFETs

I_D: MOSFET drain current

t_R, t_f: rise and fall time of the switching waveform

F_{sw}: switching frequency

Equation 5.2 indicates that switching loss increases linearly with input voltage and switching frequency.

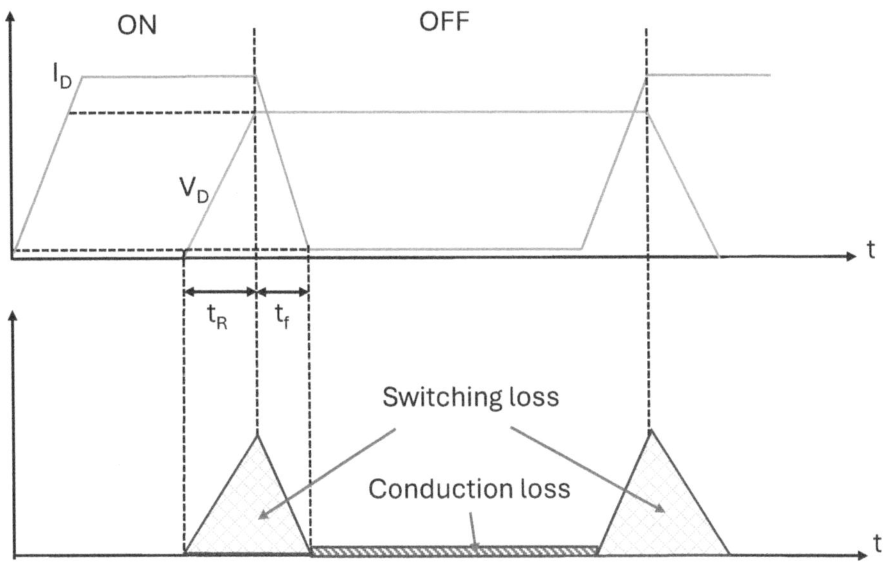

FIGURE 5.4 MOSFET switching waveforms, switching loss, and conduction loss.

For traction inverter power modules, Si IGBT and SiC MOSFET are the choices for switching devices (Goli, 2021). Loss estimation for IGBT devices is similar (Graovac, 2009). As Jenkins (2019b) revealed, loss mechanism for MOSFETs is different from IGBTs. One of the characteristics is that the MOSFET R_{on} increases with temperature, doubling the value from 25 to 125°C. IGBT R_{on} also increases with temperature but is less sensitive. In comparison, MOSFETs have significantly lower switching losses than IGBTs, making them more suitable for HF applications.

Loading conditions affect the losses and efficiencies in a traction inverter. Like a motor, an inverter tends to have lower efficiency under light load conditions since the total input power is low. The highest efficiency is normally reached around medium load. As the load keeps increasing, the loss will increase, the temperature will rise, and efficiency will drop. The second factor is thermal management. Poor thermal management can lead to higher temperatures, resulting in higher resistance and higher loss, thereby reducing efficiency. The third factor is control and gate driver design. The gate driver and control are critical to maximizing inverter efficiency. Switching loss is determined by the edges of the voltage and current waveforms during switching. A good gate driver design will enable fast charging and discharge of the gate, resulting in shorter rise time and fall time of the switching waveform. Optimum control helps to improve the motor control loop, reducing HF harmonics that do not contribute to useful output (Dearien, 2022). Of course, power loss also depends on the type of semiconductor devices, which will be discussed in a later chapter.

POWER ELECTRONIC COMPONENT THERMAL DESIGN

Power electronic system design is essentially about its power capability, meaning how much power it can handle. The power capability is closely tied to thermal management, meaning how much power it can handle without overheating, and ensuring its temperature is within the preferred range. The main design target of EV power electronic components is to achieve the best thermal design to minimize component temperature rise under maximum power. Based on Equation 1.5, this means the lowest thermal resistance and the highest efficiency. Critical component materials, including the die, die attach, and insulation, have temperature limits. In addition, they are more reliable at lower temperatures. The higher the power, the higher the power loss, the higher the thermal resistance, and the easier to reach temperature limits. Ultimately, the power capability of power electronic components is limited by the temperature limits of the materials, thermal design, and component efficiency.

The objective of thermal design is to dissipate the heat away effectively and ensure the temperature is within the limits. Most of the power loss is concentrated in the power semiconductor devices with a die area of just a few mm^2. It is critical to ensure this highly concentrated heat will be effectively dissipated.

In addition, the temperature of components with a low temperature rating, such as the capacitor and sensors, needs to be evaluated. The temperature of components with intrinsically poor thermal structures, such as transformers and inductors, also need to be evaluated. Busbars and circuits that conduct hundreds of amperes of current also need to be checked.

Power density of current EV traction inverters is around 10–20 kW/L, or even higher at 30 kW/L (Goli, 2021). Power density for OBCs is around 3.3 kW/L (Yuan, 2021). Due to the intrinsically high power density, thermal design is one dominant factor driving the EV/HEV power electronic system design, especially when power electronic components/systems are moving toward higher integration, higher power density, and yet lower cost, at an unprecedented speed.

In addition, higher temperatures also introduced long term reliability issues. Thermo-mechanical fatigue is an important topic. Lifetime assessment based on thermo-mechanical fatigue is important for the preliminary selection of power modules for EV applications. These topics will be captured in Chapter 6.

Thermal Design for Power Semiconductor Devices

Silicon-based power semiconductor devices are normally required to operate at junction temperatures under 150 or 175°C. SiC-based devices can operate at junction temperatures up to 200°C. Even if the devices can be operated at high temperatures, the operating temperature is limited by other materials such as joints and wire bonding due to long-term reliability requirements. For example, some solder joints are rated at temperatures less than 120°C. In addition, lower operating temperatures are always preferred for better thermo-mechanical reliability. The rule of thumb for semiconductor devices is that each 10–15°C temperature rise above 50°C will double the failure rate (Mohan, 2002).

Power electronic packages are designed to ensure the electrical and thermal performance of the devices. In EVs, the power flowing through the inverters typically ranges from 80 to 100 kW, with some reaching 300 kW. OBCs typically range from 3.3 to 11 kW. DC-DC converters range from 1.0 kW to 3.5 kW. There are two types of power electronic packages: discrete packages and multichip power modules. For low power applications such as DC-DC and OBCs, discrete packages are the common choice due to cost and wide availability. Discrete power electronic packages have the flexibility to accommodate different designs. In discrete power packages, normally only one device is included in a standard package, such as TO220, TO247, and DPAK. The dimensions of the package are the same regardless of the devices inside and the manufacturers. The device inside can be either transistors such as MOSFETs or diodes. For high power components used in traction inverter applications, due to higher demand for electrical performance and higher power and current, the designs (thermal and electrical) are challenging and therefore require significantly more development. Therefore, multichip power modules are the most popular choice (Jones-Jackson, 2022). To summarize, power modules offer a validated and specified solution for higher power applications such as inverters, while a discrete package enables more customization for applications including OBCs and DC-DC converters.

TO220- AND TO247-BASED DISCRETE POWER ELECTRONIC PACKAGES

In TO220 or TO247 (see Figure 5.1), the device is bonded on a Cu metal tab (or lead frame) via a solder joint die attach (see Figure 5.2). The package can be mounted to a heat sink to dissipate the heat. The circuitry inside the devices is normally connected via wire bonds to the leads. There are three leads for the MOSFET: source, drain, and gate connections. The metal tab is electrically connected to the internal circuitry, such as drains for MOSFETs. The package is encapsulated by a molding compound. TO247 is very similar to TO220 except that it is larger and can handle more power (see Figure 5.5).

In power electronic packages, thermal resistance is used to characterize the thermal performance of the package. Datasheets for the discrete power devices typically provide several values (see Figure 5.6):

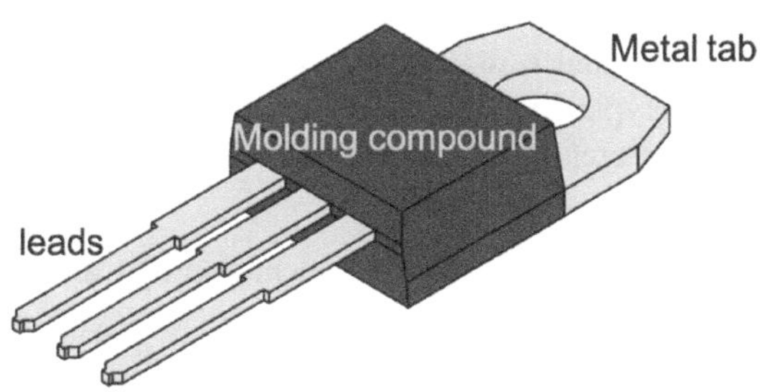

FIGURE 5.5 Front view of a TO247 (Copyright Inductiveload, permission from https://commons.wikimedia.org/wiki/File:TO-220_Front_Coloured.svg?uselang=en#Licensing).

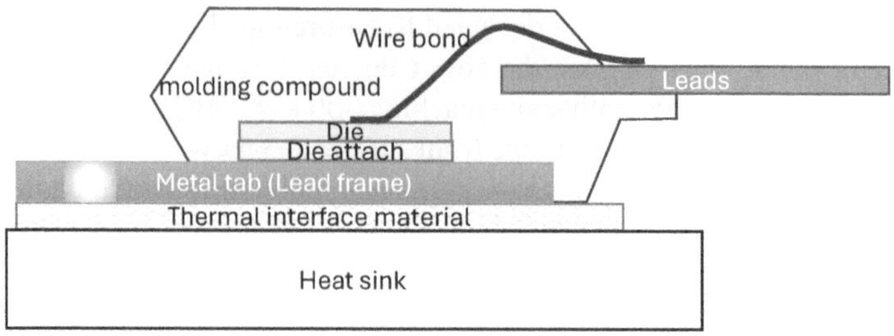

FIGURE 5.6 Cross-section view of a TO220 package attached to the heat sink, with TIM added between them.

Thermal resistance from junction to case:

$$R_{thjc} = \left(T_j - T_c\right) / P \tag{5.3}$$

Thermal resistance from junction to ambient:

$$R_{thja} = \left(T_j - T_{amb}\right) / P \tag{5.4}$$

Here,

T_j: junction temperature (°C)
T_c: case temperature (°C). Case temperature is the temperature of the metal tab.
T_{amb}: ambient temperature
P: Power loss from the device (W)

One example is IPP024N08NF2S (Infineon, 2022). In the datasheet, three thermal resistances are provided.

Thermal resistance, junction to case R_{thjc}: 0.7°C/W
Thermal resistance, junction to ambient R_{thja}: 40°C/W, when devices are mounted on a 40 mm×40 mm ×1.5 mm epoxy printed circuit board (PCB) flame retardant 4 (FR4) with a 6 cm² copper area.
Thermal resistance, junction to ambient, R_{thja}: 62°C/W with minimal footprint.

The junction-to-case thermal resistance is normally much smaller than the case-to-ambient resistance. Typically, R_{thjc} varies from 0.5 to 4°C/W (Wikipedia, 2024). R_{thjc} depends on the die area inside the package. The larger the die area, the lower R_{thjc}.

The final R_{thja} from junction to ambient depends on R_{thjc} and R_{thca}, the resistance from the case (metal tab) to ambient.

$$R_{thja} = R_{thjc} + R_{thca} \tag{5.5}$$

R_{thca} depends on the type of case, ambient coupling method, and cooling method. If there is minimal footprint, R_{thca} will be high and R_{thja} will be as high as 62°C/W. Here, the maximum junction temperature is 175°C and the ambient temperature is 25°C; the maximum power loss can be estimated based on Equation 5.4.

$$P = \frac{175^o C - 25^o C}{62W/^o C} = 2.82W$$

But if the package is mounted on 40 mm×40 mm×1.5 mm epoxy PCB FR4 with 6 cm² copper area, heat dissipation is better and R_{thja} will reduce to 40°C/W based on the datasheet. Now the maximum power dissipation allowed is:

$$P = \frac{175^o C - 25^o C}{40W/^o C} = 3.75W$$

For the same package, if R_{thja} can be further reduced with even better thermal design, the package can handle significantly higher power. Improvements in thermal management will enable the package to be operated under higher power and higher power density. A typical way for a TO package to handle more power is to attach the device to a heat sink. The metal tab of the package typically has a hole, allowing the package to be fastened to the heat sink. A clip is another option to mount the package to the heat sink. Infineon provides a detail guideline about how to assemble TO packages (Infineon, 2017). The heat sink can be cooled by either air or liquid. This will significantly reduce the thermal resistance. R_{thja} of 3°C/W for a TO220 package is achievable with this type of cooling method. With this, the power loss allowed is increased to 50 W.

$$P = \frac{175^o C - 25^o C}{3W/^o C} = 50W$$

Since the metal tab of the package is electrically active, an electrically insulated pad is required when the package is mounted on the heat sink (shown in Figure 5.2). This can be a layer of TIM or the insulation layer in the PCB or IMS. The thermal conductivity of those insulation materials is low: 1 W/mK or less, compared to 380 W/mK for Cu. Therefore, it contributes to a large portion of thermal resistance. To minimize thermal resistance, the insulation layer needs to be as thin as possible, and 0.1–0.2 mm is a typical value.

DPAK/D2PAK Discrete Power Electronic Packages

In addition to the TO220/TO247 package, another type is the surface mount package, such as DPAK/D2PAK. Like the TO247, DPAK/D2PAK are plastic packages with a standard footprint. Unlike the TO247, both lead frame and leads for DPAK are intended to be directly soldered on the substrate. The substrate is either an IMS or an FR4-based PCB substrate. Inside the package, the power device (die) is directly soldered onto a Cu metal tab (or lead frame), the same as the TO247. The Cu lead frame is active and will be the drain if the device is a MOSFET (see Figure 5.3, Figure 5.4, Figure 5.7 and Figure 5.8).

Here R_{thja} is the junction-to-ambient resistance, which is defined the same way as TO247. The lower the R_{thja}, the more power (power loss) can be applied without exceeding the junction temperature limit. DPAK are typically cooled from the bottom on the substrate side. Correspondingly, the main heat dissipation path is

Die (junction) – die attach – lead frame – solder – substrate (PCB or IMS) – ambient.

R_{thja} is affected by the Cu metal pad on the PCB it is soldered to. Figure 5.9 shows the relationship between the pad area and R_{thja}. Using NTB60N06 (On Semiconductor, 2019) as an example, based on the datasheet, R_{thja} is 62.5°C/W when the surface is

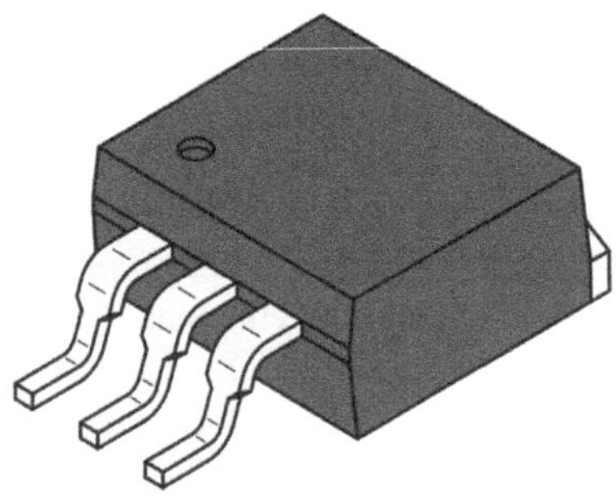

FIGURE 5.7 Front view of a DPAK package (Inductiveload, Wikimedia Commons, permission from https://commons.wikimedia.org/wiki/File:TO-263AA_Back_Top.svg).

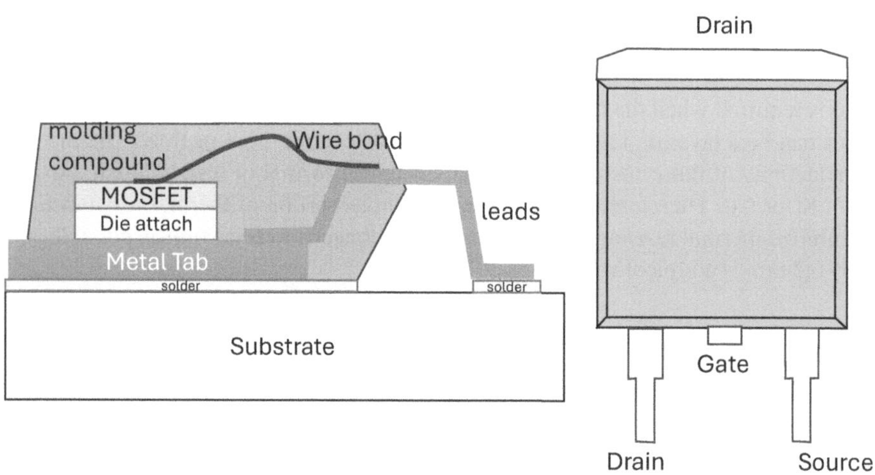

FIGURE 5.8 A MOSFET with a D2PAK package. Left: cross-section view assembled on the substrate; Right: top view.

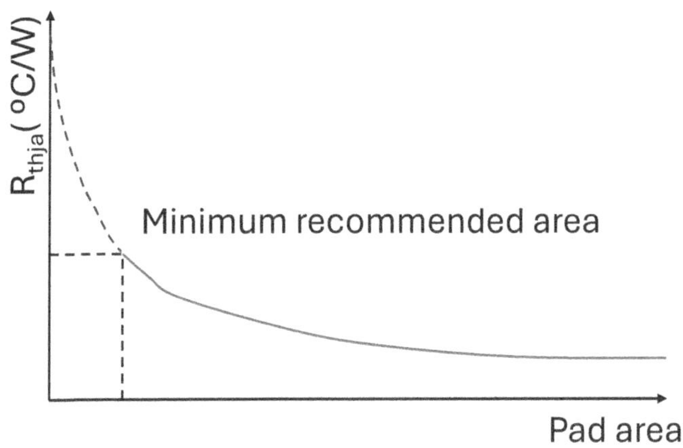

FIGURE 5.9 R_{thja} for D2PaK changes with PCB copper pad.

mounted on the minimum recommended footprint size of 0.412 in^2. Assuming T_{amb} =25°C and T_j=175°C, based on Equation 5.4, the total power dissipation allowed is P=(175°C–25°C)/(62.5 W/°C)=2.4 W.

As pad area increases, thermal resistance will reduce. The reduction effect will diminish as the area increases to a certain value (Gulino, 2003).

DPAK are commonly rated for 50 W power dissipation, while D2PAK are rated at 100 W. This is achieved when the package is mounted properly with an adequate PCB pad and heat sinking.

One major factor limiting the thermal performance of D2PAK is the thermal resistance of the PCB. Thermal resistance of the PCB depends on the thickness, layout, and number of Cu layers and FR4 layers. The very low thermal conductivity of FR4 (0.8–1.1 W/mK) is the major factor limiting thermal resistance of the PCB. There are several improvement options (Mohan, 2019). One option is thermal vias. Thermal vias are plated through holes in the PCB or IMS board that can be filled with conductive epoxy and plated over. They are typically placed directly under the components. Thermal vias have much higher thermal conductivity than FR4s, providing a high thermal conductivity path. Another option is to replace the PCB with an IMS. An IMS is an alternative substrate solution to handle more power (Hammerschmidt, 2016). It is typically a three-layer substrate. The top layer is the circuit layer made of copper. The circuit layer has a thickness that varies from 0.6 to 10 oz (1 oz is 34.8 μm in PCB Cu trace thickness). The middle layer is the dielectric layer made of polymer. The dielectric layer has thermal conductivity around 1–7 W/mK. This is much better than FR4 (~0.3 W/mK). At the same time, the thickness of the dielectric layer is 40–200 μm. It is preferred to be as thin as possible to achieve optimum thermal performance without sacrificing electrical insulation requirements. The bottom is the base layer made of either aluminum or copper. The typical base layer is a 1.57 mm thick aluminum (Bergquist, 2020). New IMSwTPG provided by TCLAD can even provide thermal performance comparable to direct bond copper (DBC) substrates (Taylor, 2023) (see Figure 5.10).

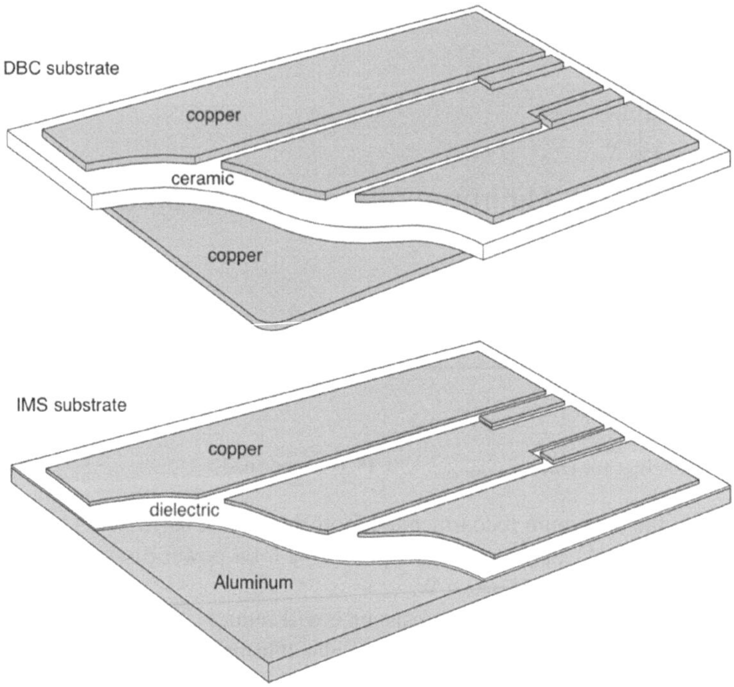

FIGURE 5.10 DBC substrate and IMS substrate (CyrilB~commonswiki, licensed under the Creative Commons Attribution-Share Alike 3.0 Unported, from https://commons.wikimedia.org/wiki/File:DBC.svg).

One advantage of D2PAK over TO247 is that it has shorter leads and can be assembled closely with other surface mount PCB components such as capacitors, achieving better inductance than TO247.

Multichip Power Module

Multichip power module contains multiple semiconductor devices within a single module. An EV traction inverter normally uses a power module. It includes several IGBTs and diodes or MOSFETs based on typical inverter topologies, which are either half bridge or three-phase full bridge (see Figure 5.8). Three-phase full bridge is also called a six-pack.

The advantages of power modules are that they offer a more compact and integrated design by packaging more devices in a module; they normally provide better thermal performance due to the use of ceramic-based substrates; and they enable better electrical performance, such as low inductance (see Figure 5.11).

The power module stack is designed to achieve the lowest thermal resistance R_{thjf} possible. For a typical power module (see Figure 5.12), the die is attached to the substrate via the die attach. Substrates include active circuitry and necessary electrical

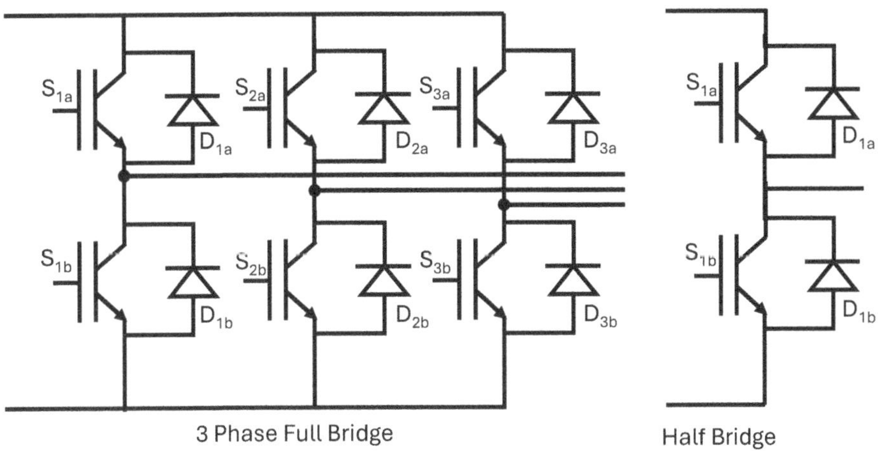

FIGURE 5.11 Topology of three-phase full bridge and half bridge. S_{1a}, S_{3b} are switches, D_{1a}, …, D_{3b} are diodes.

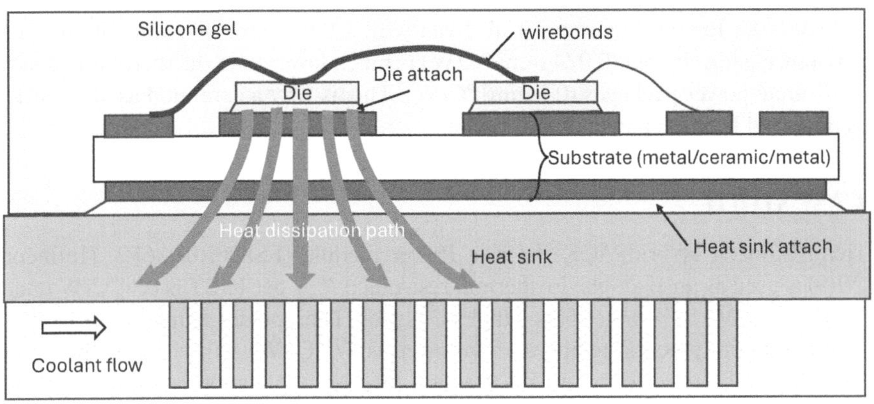

FIGURE 5.12 Cross-section view of a liquid-cooled power module; the majority of heat is dissipated from the die toward the coolant.

insulation. Then the substrate is attached to the heat sink by another joint. In the past, thermal grease was used as the TIM between the substrate/baseplate and the heat sink. Nowadays, solder joints (~50 W/mK) are becoming the mainstream choice to achieve better thermal performance.

The heat dissipates from the chips through the substrate to the heat sink. In the end, the heat is taken away by the coolant under the heat sink. In the EV traction inverter, liquid cooling with water glycol is the default option to cool the power module (Goli, 2021). Thermal resistance of the power module is characterized as R_{thjf}, thermal resistance from junction to fluid (coolant).

At steady state, the junction temperature T_j of the device is

$$T_j = T_{coolant} + PR_{thjf} \tag{5.6}$$

P: power loss
$T_{coolant}$: coolant temperature

The heat is mostly dissipated through the stack via conduction and then to the coolant via convection (see Figure 5.9). Therefore, thermal resistance R_{thjf} includes two parts: resistance due to conduction from the die to the heat sink and resistance due to convection from the heat sink to the coolant.

Resistance due to conduction:

$$R_{th\_conduction} = \frac{thickness}{KA} \tag{5.7}$$

A: conduction area

The thinner the layer, the higher the thermal conductivity (K), the larger the area (A), the lower the thermal resistance. Table 5.1 lists thermal conductivity (K) and thickness (t) for a typical power module. Ceramic has relatively high thermal conductivity (~26 W/mK) for insulation material. Even with a 0.635 mm thickness, its thermal resistance per unit area (0.024 mm² °C/W) is much lower than the thermal pad used in discrete power packages (0.1 mm² °C/W). This is why a ceramic-based substrate is preferred for high power applications.

CASE STUDY

Here, Infineon HybridPACK™ Drive Power module (FS820R08A6P2) (Infineon, 2019) is used as an example. Its thermal resistance R_{thjf} per IGBT is 0.14°C/W under 75°C, 10 LPM, 50% water/50% ethylene glycol. The coolant temperature is 75°C. The maximum junction temperature allowed is 175°C. We will check if the power module can be used for a 150 kW traction inverter application, assuming inverter efficiency is 98%.

R_{thjf} per IGBT is 0.14°C/W.

TABLE 5.1
Thermal Properties and Thickness for a Typical Power Module

	Material	Thermal Conductivity (W/mK)	Thickness (mm)
Die	Si	~130	0.07
Die Attach	Solder	~60	0.1
Substrate	$Cu/Al_2O_3/Cu$	385/~26/385	0.3/0.635/0.3
Substrate Attach	Solder	~60	0.2
Heat Sink	Cu	385	3

For a 150 kW inverter with 98% efficiency, the total power loss for the power module is:

$$P = 150kW \times (1 - 98\%) = 3000W$$

Assuming loss is only generated by IGBT and there are 6 IGBTs per module, loss per IGBT is

$$3000W / 6 = 500W$$

Based on Equation 5.6, the junction temperature is determined.

$$T_j = 500W \times 0.14°C / W + 75°C = 145°C < 175°C$$

Therefore, the power module can be used for a 150 kW traction inverter.

R_{thjf} (0.14 °C/W) for a traction inverter power module is one order lower than that of similar liquid cooled discrete packages with a thermal pad (single digit °C/W) (Future Electronics, 2020). In some very high-power applications, increasing numbers of devices in parallel is another option to further reduce both loss per die and thermal resistance.

Because heat loss is concentrated in a small area, heat spreading plays an important role (see Figure 5.12). Heat spreading increases the heat dissipation area. Heat spreading depends on the thickness and thermal conductivity of each layer as well as the heat sink and coolant. For example, a thicker layer will help the spreading, but it increases the conduction path at the same time. As the area increases beyond a certain value, the spreading effect diminishes.

Choosing material with high thermal conductivity is another option to reduce thermal resistance. As discussed earlier, one bottleneck for power module thermal performance lies in the low thermal conductivity insulation layer in the substrate. Electrical insulation is required to electrically isolate the active circuits on the substrate from a heat sink. Typically, high electrical insulations are accompanied by low thermal conductivity. In current inverter power modules, DBC is the most used substrate. It includes a ceramic layer sandwiched between two copper layers. Ceramic can be Al_2O_3 (24–26 W/mK), Si_3N_4(55 W/mK), and AlN (180–230 W/mK). These ceramics have much better thermal conductivity than other typical insulation materials. Al_2O_3 is the cheapest. AlN is the most expensive but most effective. Thinner ceramics are always preferred thermally. The drawback is that it compromises mechanical strength. Both aluminum oxide and nitride are ceramic-like materials that are hard and brittle, so insulators made from them must be relatively thick. Typically, 0.3 mm is a typical insulation thickness (Chen, 2017; Rogers Corporation, 2025).

For multichip modules, cross-coupling is an important factor that affects junction temperature. Cross-coupling happens when two or more chips share the same substrate and are close to each other; the temperature rise of one chip is affected by the other when both are turned on. The impact depends on the power loss, the distance between chips, and the structures underneath. Generally, a shorter distance between chips in inverter modules is desired to achieve low inductance and a smaller footprint. However, if the distance is too close, thermal resistance can increase significantly due to cross conduction.

Heat sink design is very important to cool the inverter power module. The design of the heat sink should consider factors such as weight, size, and manufacturability. Cu is the most common heat sink material for its excellent thermal conductivity. However, it is heavy, expensive, and needs to be plated with nickel to eliminate galvanic corrosion when other Al parts exist in the cooling loop. Al heat sinks, on the other hand, are gaining popularity in EV traction inverters, although they have much lower thermal conductivity. However, this does not significantly affect the overall thermal resistance (Zheng, 2016). Also, they are lighter and cheaper, and do not require additional plating.

In power modules, the pin fin heat sink is one popular choice (Nakatsu, 2014). The higher the fins' density, the larger the cooling area, and the lower the thermal resistances. However, this effect diminishes as the fins becoming more densely populated. On the other hand, more fins create more resistance to coolant flow and result in higher pressure drop. For liquid cooling, the flow rate is important. A higher flow rate leads to better thermal performance, but the rate of gains reduces after a certain rate (Infineon, 2019). In addition, higher flow rate is associated with a higher pressure drop. Bypass is another important factor that needs to be considered in heat sink design. The coolant bypass flow occurs between either at the tips of the pins or next to the flow channel (Horiuchi et al., 2013). Bypass will reduce cooling efficiency.

One key feature of the traction inverter power module is its excellent thermal performance and capability to handle high power for traction applications, which is critical for ensuring reliable operation, optimizing performance, and extending the lifespan of the EV traction system.

Transient Performance

Transient thermal response of components describes how quickly the component temperature changes when power is applied or removed. During a transient event such as a power or current surge, if the heat cannot be dissipated fast enough, the power device junction temperature can exceed the limit in a short time, and the device will fail.

Transient thermal behavior of the power module affects peak power and peak current performance of power electronic components. The traction inverter typically has a much smaller heat capacity than motors due to its thermal mass. This leads to quicker temperature changes when power is applied or removed. For liquid-cooled traction inverter power modules, it typically takes less than a few seconds (<10 seconds) for the module to reach steady state (Infineon, 2019). It will be even shorter for liquid-cooled discrete power packages. As discussed in Chapter 2, peak power and peak current typically last a few seconds for EV applications. The EV inverter power module can be close to its steady state when peak power or peak current occurs.

Most power electronic components are rated for peak power (Infineon Team, 2023). The peak power of the traction inverter is higher than its continuous power capability. One reason is that the junction temperature limit of the power devices under continuous conditions is different from the limit under peak conditions. Using Infineon Hybrid pack drive module (Infineon, 2019) as an example, maximum allowed T_j is 150°C under continuous conditions. However, it increases to 175°C 'for 10 seconds within a period of 30 seconds, occurring a maximum 3000 times over the lifetime.' For inverters, typically the maximum power loss occurs at maximum current, which aligns with peak torque.

TREND

The trend for power electronic components packaging has always been higher power density, better performance, and higher integration. Robust thermal performance is a key enabler. The advances in thermal performance in both discrete packages and power modules are discussed next.

Discrete Power Electronic Packages

To achieve higher power density, ceramic-based substrate copper is replacing lead frames in discrete packages, eliminating the need for additional TIM. One example is the TO247-based SiC MOSFET used in the Tesla Model 3 traction inverter (Jenkins, 2020; Barbarini, 2018; PntPower, 2015). Electrical insulation is provided by ceramic in the substrate. Therefore, the TO247 can be directly attached to Al heat sink without additional TIM. As a result, the discrete package-based inverter was able to achieve a power density that is comparable or even better than some of the power module-based EV traction inverters, yet with lower cost. Similarly, Littlefuse developed the ISO247 package which has the same outline dimensions as TO247, but uses a ceramic substrate to minimize thermal resistance. This has resulted in significantly lower SiC MOSFET chip temperature (Perraud, 2023). It should be noted that the leads on the ISO247 package need to be kept short to ensure sufficient current capacity. Otherwise, there may be tens of amperes of current going through the leads without any effective heat dissipation path.

Wire bonds inside the package can also be improved with emerging package technology, where Cu clips are used to replace wire bonds. A larger cross-sectional area has significantly reduced R_{dson} compared to Al wire. Therefore, lower power loss and higher current capacity can be achieved with a smaller footprint (Electronics360, 2018).

Double Side Cooling Power Module

Most EV power inverter modules are cooled through one direction (shown in Figure 5.12): die to substrate to heat sink. Therefore, the heat will dissipate only on one side. Cooling from both sides has the potential to double the heat dissipation area, reduce the thermal resistances, and significantly increase the heat dissipation capability of the module. Double side cooling power modules are currently used in several EV traction inverters, including modules from Delphi, Hitachi, and Denso (Liu, 2022). The stack up is illustrated in Figure 5.13.

In Delphi Viper power module, the power devices are sandwiched between two AlN-based direct bond aluminums (DBAs) and encapsulated in the package (Barbarini, 2019). They are coupled to the heat sink from the top and bottom via the TIM (Anwar, 2015). Delphi's Viper double side cooling module was used in the Chevrolet Volt. It resulted in 25% higher power density because of better thermal performance, and achieved range improvements of 25% (Liu, 2022; Delphi, 2023). At the same time, the module achieved a 10% reduction in cost, a 40% reduction in mass, and is 30% smaller in volume.

Hitachi's double side cooling model is different. It includes a heat sink. The devices inside the power module are soldered to a Cu lead frame on both sides, with a spacer added to accommodate the height difference between devices within the same package. The Cu lead frame on both sides is then attached to Al heat sink via an isolation sheet. The package was encapsulated. As reported by Hitachi, the module has resulted in a 35% improvement in thermal resistance over a single-sided module in an IGBT module of the same size. This leads to a 75% higher current density (Nakatsu, 2014). The module also achieved 30% better in terms of current capacity for the same chip size (Hitachi, 1995/2024). The module was used in the Audi Etron (Hitachi, 2019). 'The power density output of this inverter has been increased 160% compared to the previous generation of inverters.'

In Denso's double side power module, power devices were sandwiched in between two copper lead frames. Spacers are used to accommodate differences in device thickness. The package was molded with epoxy. The module was coupled to Si_3N_4 from both sides with TIM, then to the heat sink by TIM again. The Denso double side cooling module was used in the Toyota Prius. It is about 33% smaller than that of the previous Prius and is more efficient (Denso, 2015).

Another advantage of double side cooling lies in low parasitic inductance (Liu, 2022; Lovati, 2023). The layout in double side cooling can be optimized to achieve

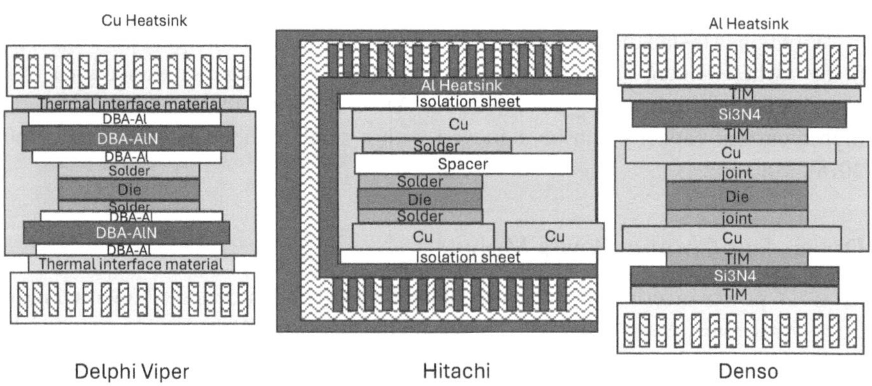

FIGURE 5.13 Double side cooling power module stack (Liu, 2022; Barbarini, 2019; Yang, 2020).

low inductance of less than 10 nH, while traditional single power modules have parasitic inductance around 15–20 nH. 'The reduction in inductance allowed the high-speed switching, and make it possible to reduce the switching loss drastically.'

Typically, double side cooling requires high precision control of the stack-up thickness; some require spacers to accommodate the differences due to thickness variation between switch and diode. Also, module to external connection design becomes more challenging.

BIBLIOGRAPHY

Anwar, M. e. (2015). Development of a power dense and environmentally robust traction power inverter for the second-generatio chevrolet VOLT extended-range EV. *2015 IEEE Energy Conversion Congress and Exposition (ECCE)*. Montreal, QC, Canada: IEEE.

Barbarini, E. E. (2018, June). *STMicroelectronics SiC Module Tesla Model3 Inverter*. From YoleGroup www.systemplus.fr: https://medias.yolegroup.com/uploads/2019/02/YOLE_SP18413-STM_SiC_Module_Tesla_Model_3_Inverter_sample.pdf

Barbarini, E. E. (2019). *Power Module Packaging: Market and Technology Trends*. From IMAPS Source Proceedings: https://www.google.com/url?sa=t&source=web&rct=j&opi=89978449&url=https://imapsource.org/api/v1/articles/67312-power-module-packaging-market-and-technology-trends.pdf&ved=2ahUKEwiV-4PUrsmJAxXEM0QIHUtaLMs4ChAWegQIExAB&usg=AOvVaw2OB-6s2IINXE0NbuZ7TvOC

Bergquist. (2020). *Thermal CLAD Compresensive Selection Guide*. From https://dm.henkel-dam.com/is/content/henkel/Bergquist%20Comprensive%20Selection%20Guide%20-%20Thermal%20Cladpdf

Biancomano, V. (2011, Jul 1). *Inverters Fight for Traction*. From Electronic Design: https://www.electronicdesign.com/technologies/power/power-supply/power-electronics-systems/article/21193481/inverters-fight-for-traction

Blaabjerg, F. (2021). Reliability of power electronic systems for EV/HEV applications. *Proceedings of the IEEE*, *109*(6), 1060–1076.

Chen, C. E. (2017). A review of SiC power module packaging: Layout, material system and integration. *CPSS Transactions on Power Electronics and Applications*, *3*(3).

Chowdhury, S. E. (2020). High-energy density capacitors for electric vehicle traction inverters. *IEEE Transportation Electrification Conference & Expo (ITEC)*. Chicago, IL, USA: IEEE.

Dearien, A. E. (2022, Sep). *Traction Inverters – A Driving Force Behind Vehicle Electrification*. From Texas Instrument: https://www.ti.com/lit/wp/slyy217/slyy217.pdf?ts=1730645030249&ref_url=https%253A%252F%252Fwww.google.com%252F

Delphi. (2023). *Traction Power Inverter Module*. From Delphi Power Electronics: https://www.delphiautoparts.com/products-technologies/vehicle-electronics-engine-management/power-electrics/traction-power-inverter-module#Product-Information-1

Denso. (2015, Oct 2015). *DENSO Develops Products Used in Toyota's New Prius*. From Denso News: https://www.denso.com/global/en/news/newsroom/2015/20151029-01?

Donaldson, P. (2021). *Cell-to-Pack Batteries*. From https://www.emobility-engineering.com/cell-to-pack-batteries/

Electronics360. (2018, Apr 5). *The Benefits of Copper Clip Over Wire Bond Packaging*. From https://electronics360.globalspec.com/article/11548/the-benefits-of-copper-clip-over-wire-bond-packaging

Future Electronics. (2020, Dec 17). *Infineon: HybridPACK™ Drive | Power Modules for Main Inverter Applications*. From Youtube: https://www.youtube.com/watch?v=73SHs3Fa5z4

Goli, C. E. (2021). Review of recent trends in design of traction inverters for electric vehicle applications. *021 IEEE 12th International Symposium on Power Electronics for Distributed Generation Systems (PEDG)* (pp. 1–6). Chicago, IL, USA: IEEE.

Graovac, D. E. (2009, Jan). *IGBT Power Losses Calculation Using the Data-Sheet Parameters.* From Infineon Application Note, V 1.1: https://community.infineon.com/gfawx74859/attachments/gfawx74859/mosfetsisic/5344/1/IGBT%20Power%20Losses%20Calculation%20using%20the%20Data%20Sheet%20Parameters-1.pdf

Gulino, R. (2003, Jun). *AN1703 Application Note: Guidelines for Using St's Mosfet SMD Packages.* From https://www.st.com/resource/en/application_note/an1703-guidelines-for-using-sts-mosfet-smd-packages-stmicroelectronics.pdf

Hammerschmidt, C. (2016, Oct 12). *How Automotive Applications Benefit from Advanced MOSFET Packaging.* From EENews Europe: https://www.eenewseurope.com/en/how-automotive-applications-benefit-from-advanced-mosfet-packaging-2/

Hitachi. (1995/2024). *Hybrid Inverter.* From https://www.hitachi-automotive.us/Products/oem/EPS/HEV/Inverter/index.htm

Hitachi. (2019, May 17). *Hitachi Automotive Systems' EV Inverter Adopted for the E-Tron, Audi's First Mass Production Electric Vehicle.* From Automotiveworld: https://www.automotiveworld.com/news-releases/hitachi-automotive-systems-ev-inverter-adopted-for-the-e-tron-audis-first-mass-production-electric-vehicle/

Horiuchi, L., et al. (2013). Experimental verification of model for liquid-cooled staggered pin fin heat sinks with top bypass flow. *Journal of Energy and Power Engineering, 7,* 1487–1495.

Husain, I. (2021). Electric drive technology trends, challenges, and opportunities for future electric vehicles. *IEEE, 109*(6), 1039–1059.

Infineon. (2017, Feb 10). *TO-247PLUS Description of the Packages and Assembly Guidelines.* From www.infineon.com Application Note: https://www.infineon.com/dgdl/Infineon-Discrete_IGBT_in_TO-247PLUS-AN-v02_00-EN.pdf?fileId=5546d46249cd10140149e0c7fe9d56c7

Infineon. (2019, Oct 09). *HybridPACKTM FS820R08A6P2 Drive Module Data Sheet.* From https://www.infineon.com/dgdl/Infineon-FS820R08A6P2-DataSheet-v03_30-EN.pdf?fileId=5546d4626cb27db2016cdda3fdba2f5e

Infineon. (2022, Jun 15). *Infineon IPP024N08NF2S datasheet Rev. 2.1.* From Infineon: https://www.infineon.com/dgdl/Infineon-IPP024N08NF2S-DataSheet-v02_00-EN.pdf?fileId=5546d46276c4f5350176f70288ae1698

Infineon. (2024). *High-Voltage DC-DC Converter for Electric Vehicles.* From https://www.infineon.com/cms/en/applications/automotive/electric-drive-train/ev-hev-power-conversion/hv-lv-dc-dc-converter/?adgroupid=146028164364&gad_source=1&gbraid=0AAAAADpmf9eJIJX2GNK_t7QuJwuSHeQ7D&gclid=Cj0KCQiAgdC6BhCgARIsAPWNWH0xVwJ0kpUuvGqLc1e3U6x

Infineon Team. (2023, Apr 17). *Electronics 101: Power Losses and Thermal Considerations.* From Infineon Developer Community: https://community.infineon.com/t5/Knowledge-Base-Articles/Electronics-101-Power-Losses-and-Thermal-Considerations/ta-p/412455#

Jenkins, J. (2019a, August 12). *A Closer Look at Current Sensors in EVs.* From Charged Electric Vehicles Magazine: https://chargedevs.com/features/a-closer-look-at-current-sensors-in-evs/

Jenkins, J. (2019b, May 10). *A Closer Look at the Losses in Power Semiconductors.* From Charged EV: https://chargedevs.com/features/a-closer-look-at-the-losses-in-power-semiconductors/

Jenkins, J. (2020, Mar 3). *Here's Why Tesla Transitioned to a Semi-Custom Power Module Design in Model 3 Inverter.* From Charged Electric Vehicle Magazine: https://

chargedevs.com/features/heres-why-tesla-transitioned-to-a-semi-custom-power-module-design-in-model-3-inverter/

Jones-Jackson, S. E. (2022, Jun). Overview of Current Thermal Management Of Automotive Power Electronics For Traction Purposes And Future Directions. *IEEE Transactions on Transportation Electrification*, *8*(2), 2412–2428.

Keim, R. (2024, Apr 28). *Introduction to MOSFET Switching Losses.* From All About Circuits: https://www.allaboutcircuits.com/technical-articles/introduction-to-mosfet-switching-losses/

Liu, M. E. (2022). *Comprehensive Review and State of Development of Double-Sided Cooled Package Technology for Automotive Power Modules.* From https://ieeexplore.ieee.org/stamp/stamp.jsp?arnumber=9769770

Lovati, S. (2023, Jun 13). *Double--Sided Cooled Architectures for Traction Drive Power Modules.* From https://www.powerelectronicsnews.com/double-sided-cooled-architectures-for-traction-drive-power-modules/

Mainka, K. E. (2011). *Counting Methods for Lifetime Calculation of Power Modules.* From Power Electronics Europe, Issue 6 2011: https://www.power-mag.com/pdf/feature_pdf/1319730265_Infineon0611_Layout_1.pdf

McDonald, B. (2022). Addressing EMI conducted emissions challenges in electric vehicles with GaN-based OBCs. *TI Analog Design Journal,* https://www.ti.com/lit/an/slyt831/slyt831.pdf?ts=1755464961165

Mohan, N. E. (2002). *Power Electronics: Converters, Applications, and Design* (3rd ed.). Hoboken, NJ: Wiley.

Mohan, S. (2019, Jun). *Texas Instruments Application Report: Thermal Comparison of FR-4 and Insulated Metal Substrate PCB for GaN Inverter.* From https://www.ti.com/lit/an/tida030/tida030.pdf

Nakatsu, K. E. (2014). A novel direct water and double-sided cooled power module for HEV/EV inverter. *2014 International Conference on Electronics Packaging (ICEP).* Toyama, Japan: IEEE.

On Semiconductor. (2019, May). *NTB60N06/D Datasheet.* From https://www.onsemi.com/pdf/datasheet/ntb60n06-d.pdf

Panasonic. (2023, Sep 22). *What Is a DC/DC Converter in an Electric Vehicle (EV)? - Device That Converts High Voltage into Low Voltage.* From https://industrial.panasonic.com/ww/ss/technical/ap7

Perraud, F. (2023, Aug 13). *SMPD: An Advanced Isolated Package To Keep the SiC MOSFET Chip Up to 75°C Cooler.* From EE Power: https://eepower.com/industry-articles/smpd-an-advanced-isolated-package-to-keep-the-sic-mosfet-chip-up-to-75c-cooler/#

PntPower. (2015, Sep 15). *In a Tesla Model S, There Is No IGBT Packaging Trick.* From PntPower.com: https://www.pntpower.com/on-tesla-electric-vehicles-semiconductor-packaging/

Rogers Corporation. (2025). *Curamik Ceramic Substrates Product Information & Data Sheet.* From https://www.rogerscorp.com/-/media/project/rogerscorp/documents/advanced-electronics-solutions/english/product-information/curamik-ceramic-substrates-product-information-and-data-sheet.pdf

ROHM CO., LTD. (1997–2024). *Power Stage Topology Selection - ON Board Charger for EV (OBC).* From https://www.rohm.com/topology/on-board-charger

ROHM Semiconductor. (2016, 10). *Switching Regulator IC Series: Calculation of Power Loss (Synchronous).* From RoHM Semiconductors Application Note: https://fscdn.rohm.com/en/products/databook/applinote/ic/power/switching_regulator/power_loss_appli-e.pdf

Skyworks. (2024). *Electric Vehicle Main DC-DC Converte.* From https://www.skyworksinc.com/Application-Pages/Automotive-EV-Main-DC-DC-Converter

Taylor, S. (2023). *Insulated Metal Substrate (IMS) with Embedded Thermal Pyrolytic Graphite (TPG), IMSwTPG, for Wide-Bandgap Semiconductor Power Modules.* From https://www.tclad.com/files/uploads/Technical%20Brief/TCLAD-%20TPG%20white%20paper-5-18-2023.pdf

Tesla. (2024). *Power Electronics Module.* From Roadster D&O: https://service.tesla.com/docs/Public/Roadster/TheoryOp/1.2.5/tabmenuframe.php-ID=DO.html

Wikipedia. (2024). *TO220.* From Wikipedia: https://en.wikipedia.org/wiki/TO-220

Wolfspeed. (2022, Jul 18). *Designing with Silicon Carbide in Bidirectional On-Board Chargers.* From Wolfspeed.com: https://www.wolfspeed.com/knowledge-center/article/designing-with-silicon-carbide-in-bidirectional-on-board-chargers/

Yang, Y. E. (2020, Sep 14). *Automotive Power Module Packaging: Current Status and Future Trends.* From IEEE: https://ieeexplore.ieee.org/stamp/stamp.jsp?arnumber=9178734

Yildirim, M. E. (2014). A survey on comparison of electric motor types and drives used for electric vehicles. *2014 16th International Power Electronics and Motion Control Conference and Exposition.* Antalya, Turkey: IEEE.

Yuan, J. (2021). A review of bidirectional on-board chargers for electric vehicles. *IEEE Access, 9*, 51501–51518.

Zheng, Y. (2016). *Some Progress in Cooling and 3D Packaging for EV/HEV Inverters.* From https://www.psma.com/sites/default/files/uploads/tech-forums-packaging/presentations/is22-some-progress-cooling-and-3d-packaging pdf

6 Power Electronic Packaging Interconnects and Thermo-Mechanical Reliability

INTRODUCTION

'You are as strong as your weakest link.' Interconnects and connections are the weakest link for power electronic packaging systems. Power electronics component interconnects include on-chip interconnects, die attach, and heat sink attach (see Figure 6.1). On-chip interconnects connect power devices to the leads or substrate. Wire bonds are the most widely used on-chip interconnects. Die attach attaches the chip (die) to the substrate. The heat sink attach is the joint between the substrate and heat sink. The solder is generally used as the die attach or heat sink attach. Steinberg (2000) pointed out, 'Most electronic failures are mechanical in nature, many of these mechanical failures occur in the component lead wires and solder joints. Extensive military testing experience over a period of many years has shown that about 80% of the electromechanical failures are due to some type of thermal condition.' Interconnect failures are the dominant fatigue failure mechanisms in power electronic packages. New interconnect technologies with better reliability performance have emerged. Sintering joints are becoming a reliable option for the die attach, heat sink attach, and even on-chip interconnects. For multichip power modules, interconnects also include power connections from lead frames to busbars and signal connections from the die to the gate drive board. Power connections could be fasteners, pressfit connections, welding connections, and ultrasonic bonding connections. Signal connection joints can be either solder joints or pressfit. Some of these interconnect technologies will be discussed as well.

Power electronic systems are normally made of different materials with different coefficients of thermal expansion (CTE). Both power on/off and temperature fluctuations introduce cyclic thermo-mechanical stress, eventually leading to interconnect fatigue failure. Wire bond fatigue and solder joint fatigue are the dominant thermo-mechanical fatigue failure mechanisms in power electronic systems, which essentially dictate power electronic systems' capability to achieve a 10 year/100,000 miles life for EV electric drive systems.

DOI: 10.1201/9781003569961-6

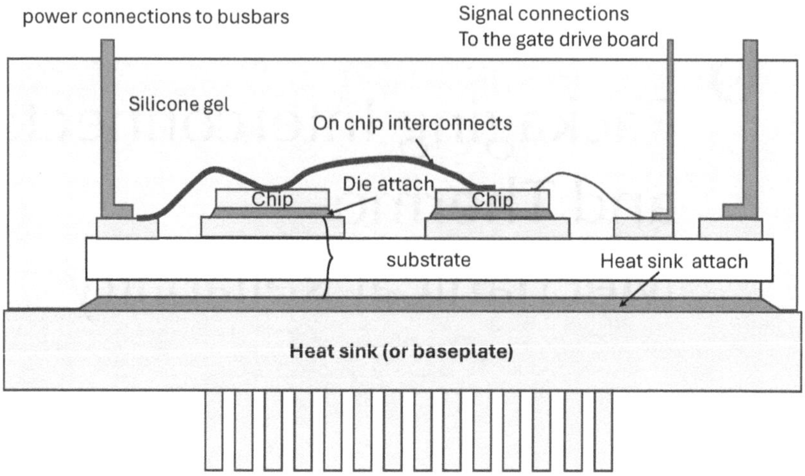

FIGURE 6.1 Schematic cross-section of a typical power module stack.

WIRE BONDING

Wire bonds are one of the most common interconnects for EV power electronic packaging. The wire bonding process is a welding process, where two metallic materials, the wire and the metallization on the substrate, are bonded together by ultrasonic energy (Harman, 1997). Al wedge wire bonding is the dominant wire bonding technology used in EV power electronic systems (see Figure 6.2).

In power electronic packages, wire bonding is used to connect the top surface of the chip (source and gate for MOSFETs) to either power or signal connections (see Figure 6.3). During the assembly process, the wire bonder presses a wire onto the chip or substrate where it is bonded. The wire bonder applies ultrasonic energy at

FIGURE 6.2 Top view of Al wire bonds in multichip power module with cover removed (Pslawinski, licensed under the Creative Commons Attribution-Share Alike 3.0 Unported license, from https://commons.wikimedia.org/wiki/File:CM600DU-24NFH.jpg).

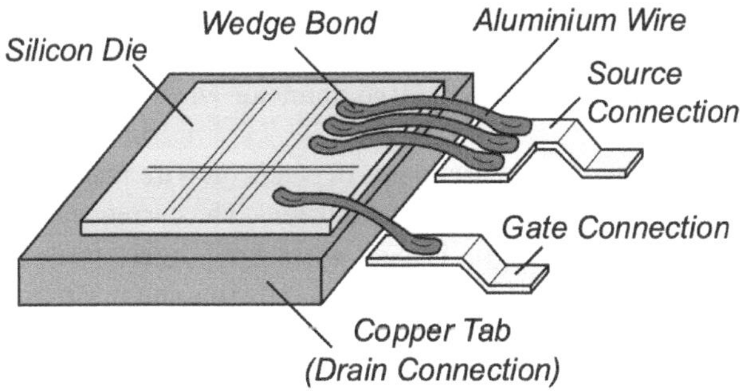

FIGURE 6.3 Al wire bonds in a power discrete package, bonding to top surface of the die (chip) on one end and to the gate connection or source connection on the other end (Inductiveload, permission granted from https://commons.wikimedia.org/w/index.php?curid=5231221).

frequencies around 100 kHz. The energy is applied to the interface between the wire and the chip (or the substrate), creating mechanical friction between the surfaces and generating localized heat. The heat softens the material at the interface and facilitates the intermingling of material from the wire and material from the chip (substrate) metallization. This is a low-temperature bonding process occurring at room temperature. The substrate or chip surface does not need to be heated. The heat generated by mechanical friction will only be applied around the interface, and the interaction (melting of metals) only occurs at the interface. Wire bonding is a very fast automatic process. Normally, it takes only a few seconds to bond a wire. Once the melt depth is reached, the process is complete, allowing the melted interface to solidify and cool, creating the bond.

Al is the most used wire bonding material in EV power electronic packaging systems. Copper wire bonding is less common, although it has better electrical and thermal conductivity. The reasons are that the raw material and process for copper wire bonding is more expensive. The process is also more difficult since copper is much harder than Al. It is also more susceptible to oxidation.

Al wire can be directly bonded to the Al surface. Wire bonding surfaces with material different from Al are typically plated with a layer of metallization. Electroless Nickel Immersion Gold (ENIG) is a popular choice. To ensure bondability and reliability of the wire bonds, the bonding surface must be non-oxidized and free of any organic contaminants, debris, and other impurities. Cleaning before wire bonding is very critical (Khan, 2018). Another key process requirement is a stable bond surface. A fixture may be needed to hold the substrate rigid. During the bonding process, ultrasonic energy is applied under pressure. Excessive vibration is likely to appear if the bond surface is not hold rigid. This introduces harmonics that may create resonance with the substrate surface, resulting in excessive vibrations. Excessive vibrations of the substrate surface can negatively interact with the bond formation, resulting in weak or no-stick bonds.

Based on the shape of its bonding tool, there is ball bonding and wedge bonding. Here, only wedge bonding is covered, since wedge bonding is preferred for EV power electronic applications where large diameter wires or ribbons are needed to handle large currents.

The current capacity of the wire bonds depends on wire material, wire diameter, and length. Thicker wires can carry higher current. The current capacity of the wires is inversely related to wire length. Longer wires have higher resistance, which will reduce their current capacity. The diameter of Al wires can reach 500 μm (20 mils). The power module in the EV traction inverter uses a lot of heavy Al wire bonds with diameters up to 300–500 μm.

F&K DELVOTEC recommends that currents should not exceed about 30 A per wire if ultrasonic wire bonding is used (Alba, 2019). Here are some initial values to consider regarding Al wire current capacity:

500 μm Al wire: 20–30 A

200 μm Al wire: 10–15 A

150 μm Al wire: 5–10 A

One advantage of wire bonding is that the quality of each bonding can be automatically evaluated during the bonding process. The bonder can check the quality by performing an in-situ pull test with a preset force value. This can identify weak or insufficient bonding failures. In addition, visual inspection is normally performed. Wire bonding is a very flexible power electronic interconnect. It is a proven mature technology with high reliability and a high yield process.

Ribbon bonding is another on-chip interconnect choice for high-power applications. Compared to wire bonds, ribbon bonds have a larger cross-section area per wire. As stated in (Inseto, 2024), one ribbon bond of 1 mm×100 μm can replace eight wire bonds of 125 μm. Instead of bonding several wire bonds, now only one ribbon bond is needed. This significantly reduces process time. Another advantage of ribbon bonds is that they offer a large surface area in proportion to their cross-sectional area. For power electronic applications where switching frequency is at least a few kHz, this is preferred to reduce skin effect and lower the loss. Ribbon bonds share the same equipment as typical wire bonds and won't introduce additional manufacturing costs.

DIE ATTACH AND HEAT SINK ATTACH

Lead-free solder joints are typical choices for both die attach and heat sink attach. Die attach is between the die and substrate. The heat sink attach is used for multichip power modules between the substrate and heat sink. Its bonding area is much larger (see Figure 6.1).

Solder joints are the most widely used joint technology. They are normally made of lead-free alloy, Sn-based (>90 wt%) with a small addition of Ag and Cu. For example, SAC305 (Sn96.5Ag3Cu0.5) has 96.5% Sn, 3% Ag, and 0.5% Cu. SAC387 is Sn95.5Ag3.8Cu0.7, meaning it is composed of 95.5% Sn, 3% Ag, and 0.5% Cu.

During the soldering process, a thin layer of solder paste (or preform) is placed between the die and substrate (or between the substrate and heat sink). Then it goes through the reflow process. During the process, solder reflows in a molten state, forming intermetallics at the interface and creating permanent joints.

The reflow soldering process for lead-free solder with a melting point of ~220°C is described in Figure 6.4.

Stage 1. Preheat

At this stage, temperature of the assembly and components is gradually increasing to ~150°C. Recommended temperature ramp rate is relatively slow (1–3°C/second), generally <5°C/second. This allows big components to have enough time to rise to the same temperature as small components.

Stage 2. Soak

At this stage, solder temperature will further increase from ~150°C to its melting point (~220°C), solder flux is active, and the temperature of the assembly is uniform and ready for reflow.

Stage 3. Reflow

At this stage, the temperature is typically higher than the melting point, and the peak temperature can be 20–40°C higher than the melting point. Now, solder melts and intermetallics form at the interface to form the bond. This lasts 30–90 seconds.

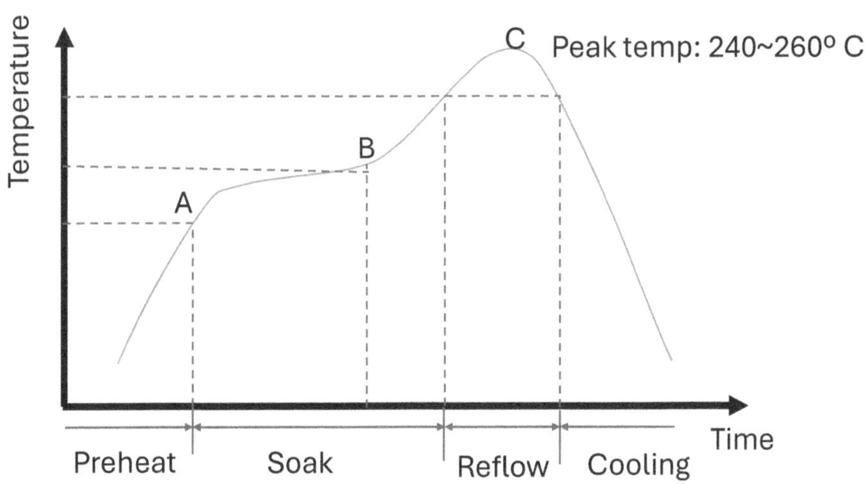

FIGURE 6.4 Reflow profile for lead-free solders.

STAGE 4. COOLING

The temperature starts to decrease at 2–4°C/second. The solder joints start to solidify, and the solder joint is formed.

An optimized reflow profile is crucial to ensure high quality and reliable solder joints. The optimum solder profile varies with the solder alloy used. It is also affected by component specifications and assembly complexity. During the soldering process, the solder joint will bond metal substrates. The indication of a good bonding is the formation of a thin layer of intermetallic (see Figure 6.5) at the interface, which is the result of metal substrates dissolving into molten solder. The type of intermetallic formed depends on substrate finishes and the solder material. Lead-free solder is usually tin based. If the substrate is copper-based, the intermetallic formed is Cu_6Sn_5 and Cu_3Sn. If the substrate is nickel-based, it will be Ni_3Sn_4. It will be Ag_3Sn for Ag-based substrates. For Sn-Ag-Cu-based lead-free solder, intermetallic (Cu-Sn or Ag-Sn) will also form in the bulk joint. These intermetallics are brittle and could have a significant effect on solder joint reliability, especially for die attach when bond line thickness is around 100–200 µm (Zheng, 2005; Ramli, 2022).

Void is always a concern for lead-free soldering processes. Voids originate from the evaporation of solder flux trapped in the solder joint during the reflow process. For power modules, voids reduce the conductive area for the current to flow and for the heat to dissipate, resulting in higher electrical resistance and thermal resistance. This negatively affects power module performance and reliability. The larger the solder joint area, the easier voids appear. Infineon's study (Infineon, 2023) has found 10% and 25% voids, with a single void underneath the center of the device, resulting in a 10% and 34% increase in thermal resistance in the TO247PLUSE reflow-solderable package, respectively. A reliable die attach or heatsink attach for power applications requires minimal voids. 'Maximum voiding percentage allowed in thermally critical zones is 5% (FET and diodes) and in mechanically critical zones is 20% (wire

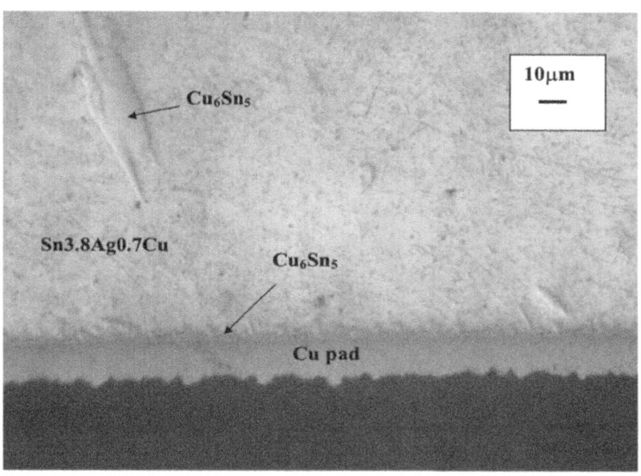

FIGURE 6.5 Optical microscopy image of Sn3.8Ag0.7Cu solder joint on OSP surface finish boards after reflow (Zheng, 2005).

bond pads),' has been reported (Hamm, 2015). Options to minimize voids include optimization of the soldering process, using solder preform and vacuum soldering.

Die attach and heat sink attach are critical to ensure effective heat dissipation from the die to the heat sink. There is a trade-off between thermal performance and thermal stress related to the dimensions of the attach. A thinner bond line results in better thermal performance. In a typical power module, the joint bond line varies from 100 to 200 μm. Joint failure is very sensitive to the bond line thickness, depending on the manufacturing process. For some processes, a thicker bond line up to a certain value is desired. One study (Guth, 2006) shows there is an optimum thickness of ~200 μm. Other studies (Morozumi, 2003; Hayashi, 2002) indicated a thickness of 200 μm or more has superior thermo-mechanical reliability for the SnAg solder die attach between the die and substrate.

Normally, the solder joint failure initiates from the edges as thermal stress is the highest at the corners (Barbagallo, 2017; Nishiura, 2007). But under power cycling, 'the start of degradation is also found below the device center' has also been reported (Lutz, 2008). The possible cause is 'During power cycling, significant temperature gradients appear between different layers. Especially using semiconductor dies with area >1 cm^2, temperature differences up to 20 K are possible between the hot center and the edge of the silicon die.' Another study (Hayashi, 2002) has investigated the effect of solder thickness on fatigue life through simulations and experiments. The study revealed that the principal factor affecting fatigue life was the minimum thickness of solder, not the average thickness. One side of the solder is thinner than the other side. This tilt results in a shorter fatigue life. The root cause of the tilt is 'when die moves and (joint) tilts in the reflow process due to the unbalanced capillary force induced by the liquidus solder' (Wang, 2016). The solution 'to improve the fatigue life of the solder joint is to prevent the tilt of the substrate, and to control the solder thickness in high-precision level.' One solution proposed (Hayashi, 2002) is to add small wire bumps near the joint corners to prevent tilt.

THERMO-MECHANICAL FATIGUE

Extensive studies (Yang, 2011; GopiReddy, 2014; Liu, 2009) have shown that power semiconductor package thermo-mechanical fatigue failures are a major concern for power converter reliability. The main thermo-mechanical fatigue failures in high power devices occurred in wire bonds, die attach, and heat sink attach.

In power electronic packages, the power loss due to power on generated a large amount of heat in the devices. The heat dissipated from the devices through die attach, substrate, heat sink attach, heat sink, and to the liquid coolant in the end. It resulted in different temperature changes across the package stack. Power on/off introduces temperature fluctuations across the packages. Temperature fluctuations and CTE mismatch (see Figure 6.6) between different layers result in cyclic thermal stress in the module. The cyclic stress eventually causes fatigue failure. Die attach, heat sink attach, and on-chip connections are the most likely failure locations.

Using the traction inverter power module as an example, the total power can be ~100 kW. Assuming 97% efficiency, the total power loss will be 3 kW. With six

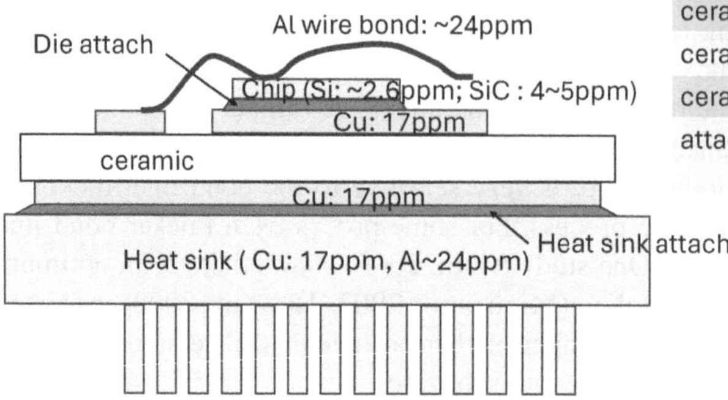

	material	CTE
ceramic	AlN	~4.7 ppm
ceramic	Si_3N_4	2.5ppm
ceramic	Al_2O_3	~7ppm
attach	SnAgCu solder	~20ppm

FIGURE 6.6 Illustration of CTE for power electronic package stack.

switches, each switch has a 500 W loss. Assuming each switch has two IGBTs and two diodes, each IGBT may have a 150 W loss and each diode may have a 100 W loss. This is significant for an IGBT die with a 10 mmx10 mm area, and the junction temperature change could be significant. With power devices switching on/off, temperature swings occur across the power devices, where the highest temperature occurs at the junction and the temperature gradient occurs across the stack depending on the on/off time. Cyclic thermal stress is produced due to temperature swings and CTE mismatch between different layers in the package. Another typical temperature swing comes from environmental temperature changes, which won't introduce a temperature gradient across the package. For wire bonds, one additional heat source is joule heating due to large current.

Die attach and heat sink attach for power electronic applications are desired to have low thermal resistance and high thermo-mechanical durability. The physics of die attach failure is the thermo-mechanical stress due to CTE mismatch between the die and the substrate. Similarly, heat sink attach stress appears due to CTE mismatch between the substrate and the heat sink. In order to minimize the stress, a better CTE match between the die and substrate is desired to improve die attach reliability. For example, both AlN and Si_3N_4 offer better CTE matches with a die than commonly used Al_2O_3. More importantly, they have better thermal performance (AlN 180 W/mK; Si_3N_4 90 W/mK) compared to Al_2O_3 (19–26 W/mK). Therefore, Si_3N_4 instead of Al_2O_3 was preferred for better power module reliability (Rogers Corp, 2024; Miric, 2015).

Under temperature cycling due to the environment, the temperature of the entire module changes at the same time, and there is little or no temperature gradient across the power module. The most significant stress occurs at the heat sink attach since it has the largest area. Therefore, heat sink attach failure is normally the dominant failure under temperature cycling (Fuji Electric Co., Ltd, n.d.; Mitsubishi Electric, 2019).

For traditional power modules, the failure typically occurs at wire bonds and die attach under short power on/off cycles (seconds on/off) (Fuji Electric Co., Ltd, n.d.; Borucki, 2018; Held, 1999; Mitsubishi Electric, 2019; Satheesh, 2014; Herrmann, 2007; Schilling, 2012) reported for accelerated power cycling (PC) with short cycle durations (<5 sec), in which wire bond fatigue is the dominant failure mode. For long duration on/off cycles in minutes, the failure locations tend to occur at the die attach and heat sink attach (Fuji Electric Co., Ltd, n.d.) (see Figure 6.7).

During power on, there is a large amount of current going through the wires. Heat is generated due to power loss of the switches and joule heating of the wires. The power loss at the wires is mostly dissipated via conduction through the thin wires to the two sides (chip or substrate) and convection to the surroundings. Neither is effective. As a result, the temperature of wire bonds can be higher than the device junction temperature. This contributes to wire bond failure. Wire bond failure is mostly a result of CTE mismatch between wire bonds (Al: ~24 ppm/°C) and the die (Si: 2.6 ppm/°C) (Borucki, 2018; Zeng, 2018). The failure mechanism is mostly wire-bond lift-off. Heel crack also happens (Borucki, 2018; Held, 1999). Another failure mechanism is grain growth (Satheesh, 2014).

POWER CYCLING TEST AND TEMPERATURE CYCLING TEST

Thermo-mechanical reliability of power electronic packages is important to ensure electric drive system performance and longevity. Typically, there are two types of reliability tests to evaluate power electronic packaging thermo-mechanical reliability: the power cycling test and thermal cycling test. Both are targeted to simulate typical applications experienced by power electronics in an accelerated way to obtain the lifetime data in a reasonable time frame, like days instead of years. The tests can be tailored according to the actual applications.

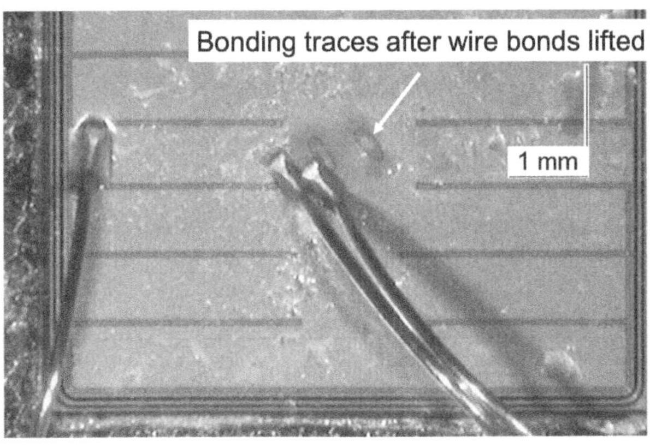

FIGURE 6.7 Wire bond lift-off failure happened to power module at the device/wire bond interface due to power cycling (Zheng, 2005).

In the thermal cycling test, the devices are not powered. A temperature swing is introduced by external temperature change such as the thermal chamber. The temperature is uniform, and there is no temperature gradient across the module. In the power cycling test, the devices are powered on/off. The temperature rise is mainly due to power loss generated in the active devices when they are switched on. Depending on the duration, there will be a temperature gradient across the module, where wire bonds and devices are the hottest. The power cycling duration per cycle can be in minutes or seconds. The objective of the power cycling test and temperature cycling test is to determine the failure mechanisms of power electronics packaging and to ensure its reliability will be sufficient for specific applications.

Power module manufacturers will have certain qualification tests to ensure power module reliability. Table 6.1 lists the power cycling and temperature cycling requirements from Fuji Electric for its power modules for automotive applications, which are more stringent than industry application.

For EV traction inverter power electronic modules, one critical design requirement is to ensure the module can survive 10 years/100,000 miles under field applications. For this purpose, lifetime analysis based on the power cycling test and temperature cycling test is performed (Siemens, 2025).

The first step of lifetime analysis is to determine the critical parameters for power module reliability. Extensive power cycling testing was conducted to understand the critical factors that affect the power module life (Bayerer, 2008; Zeng, 2018). In most practices, the junction temperature swing ΔT_{j} during cyclic change and the mean junction temperature T_{jm} are critical factors. The time duration of power on (t_{on}) during power cycling is also an important factor.

The second step is to choose the right lifetime model and to determine the parameters. Because the physics of the thermo-mechanical failure is diffusion-based, the lifetime model follows the Arrhenius term. For a specific power module, the lifetime model only varies with test conditions (Bayerer, 2008; Zeng, 2018).

$$N_f = K \cdot \Delta T_j^{\alpha} \cdot e^{\frac{E_A}{K_B T_{jm}}} \cdot t_{on}^{\beta_3} \cdot I_b^{\beta_4} \tag{6.1}$$

Here,

N_f is cycles to failure

ΔT_j is the junction temperature change over power on/off

T_{jm} is the mean junction temperature

TABLE 6.1
Temperature Cycling and Power Cycling Test Requirement (Nishiura, 2007)

	Industry Application	Automotive Application
Temperature Cycling Test	100 cycles –40–125°C	1000 cycles –40–125°C
Power Cycling Test	15,000 cycles ΔT_{j}=100°C	30,000 cycles ΔT_{j}=100°C

t_{on} is power on time
I_b is the current per chip

The rest parameters are determined by power cycling tests. Similar models exist for thermal cycling. Power cycling tests and temperature cycling tests are critical to determining the parameters for lifetime models. These tests need to be performed in an accelerated way, yet they must represent the same physics of failure.

The lifetime of the power module can be extrapolated from the equation with field condition parameters such as ΔT_j. Here, solder behavior plays an important role. Solders have melting points of ~220°C. The normal operation temperature is well above 0.5 T_{melt}. This indicates that creep is an important deformation. As a result, t_{on} is considered an important factor affecting the fatigue behavior of the joint.

Equation 6.4 has become a commonly accepted lifetime model for solder joint-based power modules, where solder is used as both die attach and heat sink attach (Rusche, 2016). Based on the results of almost ten years of +125°C operating life testing, similar Arrhenius equations have been developed by Onsemi to show the relationship between junction temperature and failure rate (Onsemi, 2023). Bosch (Schoser, 2015) has also developed a similar lifetime model based on experimental data for its power modules.

The third step for lifetime analysis is to determine the mission profile for a specific application (Mainka, 2011; Siemens Digital Industry Software, 2024; Siemens, 2025). First, different typical driving profiles are determined. A typical EV driving profile includes highway driving, start-and-stop city driving, etc. Each critical profile is transferred to motor speed, phase current, and DC link voltage in the inverter. Based on driving profile parameters combined with the electrical properties of the power module, a power loss profile for each driving profile can be determined. Next, based on the power loss profile combined with the power module thermal behavior and cooling conditions, the junction temperature change profile ΔT_j for each driving profile is determined. Eventually, the lifetime can be estimated from the junction temperature change profile based on the lifetime model.

Reliability testing, such as power cycling tests and temperature cycling tests, is still costly and time-consuming, even though they have already been accelerated. For example, based on Bayerer (2008), 20 seconds for one power cycle is assumed. For an automotive qualification test (Nishiura, 2007), 30,000 cycles of power cycling will take at least 7 days, without considering that the actual test must be stopped in between to check the degradation. In addition, multiple samples must be tested to obtain a reliable data point. To determine the parameters for lifetime model, power cycling test and temperature cycling tests have to be performed several times at different T_{jmax} and ΔT. In the end, it could take months or even longer to complete power cycling tests and temperature cycling tests for one type of power module. Once the power module changes packaging technology, dimensions, or voltage ratings, the model will likely need another round of tests.

One reason automotive power modules, especially for traction applications, are expensive is due to the development costs, where time-consuming and costly reliability tests are a major contributor. As a result, a simulation-based interconnect thermo-mechanical fatigue model has long been desired to reduce the cost. Simulation-based methods will be discussed next.

THERMO-MECHANICAL FATIGUE MODEL

During operation, power electronic packages experience temperature cycling due to either temperature fluctuation or self-heating from power dissipation. Repeated thermal cycling and CTE mismatch between devices, substrates, and other layers drive cyclic thermal stress/strain to the interconnect, eventually leading to interconnect thermo-mechanical fatigue. The thermo-mechanical fatigue model correlates thermal stress/strain to cycles to failure, where thermal stress/strain is determined by either an analytical model or finite element method (FEM) simulation.

Since cyclic stress/strain usually reaches a maximum on the edges, cracks in die attach usually initiate from the edges and propagate toward the center (see Figure 6.8). Either analytical or FEM-based analysis can be used to determine the maximum cyclic stress/strain.

For solder joint fatigue, the Coffin-Manson based model (Yang, 2011) is the most well-known basic fatigue model to correlate cyclic strain range produced by thermal cycling loading to cycles of failure (N_f).

$$\frac{\Delta\varepsilon_p}{2} = \varepsilon'_f \left(2N_f\right)^c \tag{6.2}$$

$\Delta\varepsilon_p$ is plastic strain amplitude
ε'_f is fatigue coefficient
C is fatigue exponent

For power module application, (Olsen, 1979) assumed that all the shear strain due to CTE mismatch between die and substrate ($\Delta\alpha\Delta T_j L/thickness$) is completely converted to plastic strain. However, this assumption overestimates the plastic strain and has problems capturing the temperature profile of the power module under power cycling conditions. The Engelmaier fatigue model has improved the basic Coffin-Manson model by using total shear strain. The model incorporates mean temperature

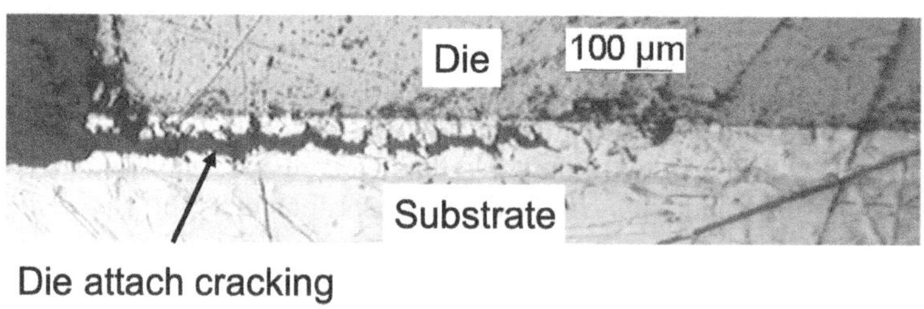

FIGURE 6.8 Die attach cracking occurred to power module due to power cycling (Zheng, 2005).

and cyclic frequency of the load. Strain range and strain energy-based models are also commonly used. Lee (2000) has provided an extensive review of those solder joint fatigue models.

One challenge in developing an accurate fatigue model is how to incorporate the elastic time-independent plasticity and time-dependent visco-plasticity (creep) properties of the solder. Typical lead-free solder readily creeps at temperatures even below room temperature. Some studies try to separate solder deformation into elastic deformation, rate-independent plastic deformation, and rate-dependent plastic deformation (creep) to estimate fatigue damage accordingly. Anand's model describes the constitutive behavior for lead free solders. In the model, both rate-independent plastic and rate-dependent creep deformations can be combined as rate-dependent inelastic deformation in unified constitutive models. Syed (2004) has developed a thermal fatigue life model for SnAgCu solder joints. The model has been correlated with a number of data points and predicts life with a discrepancy less than 25% in most cases. The life model is based on accumulated creep strain and creep strain energy density calculated from FEM simulation.

$$N_f = \left(0.0405 \cdot \epsilon_{acc}\right)^{-1} \tag{6.3}$$

$$N_f = \left(0.0014 \cdot w_{acc}\right)^{-1} \tag{6.4}$$

ϵ_{acc}: accumulated creep strain
w_{acc}: creep energy density

National Institute of Standards and Technology (n.d.) has compared extensive test results of SnAgCu lead-free solder properties and creep data. The results have found that the creep from three independent sources has first-order consistency. The study also found that the cooling rate during the soldering process appears to be an important factor, likely due to the fact that different cooling rates lead to different microstructures. This exposes the second challenge: how to consider the impact of microstructure on the material properties for the joints with bond line thickness as thin as 100 μm. Different processes will result in different microstructures.

The difference in surface finishes affects the microstructure and mechanical properties of solder joints. It is anticipated that the effect of the solder/substrate interfaces on the behavior of the solder attach will be very important, especially as the attach bond line thickness is less than 100 μm. Due to the thin bond line of die attach, surface finishes from the die and substrate sides pose a high influence on the reliability of die attach (Zheng, 2005).

Various wire bond fatigue models have been reviewed by Yang (2013) and Grams (2015). Grams (2015) provided two fatigue models. The first one is a Coffin-Manson based fatigue model. In this model, cycles of failure (N_f) are correlated to a damage parameter $\Delta\varepsilon$. The second is a crack growth model. This model correlated crack area increases per cycle to the damage parameter $\Delta\varepsilon$. In both models, $\Delta\varepsilon$ is determined by FEM simulation.

Regardless of the challenges, the simulation-based thermal fatigue model has been used in power modules to evaluate its reliability and has proven to be useful (Barbagallo, 2017; Durand, 2016; Lederer, 2022). Fatigue model is an effective design optimization tool. By simulating power modules with various material, packaging technologies, and geometries, the configuration with optimum thermo-mechanical reliability can be determined with minimal power cycling and temperature cycling tests.

To summarize, it is important to understand limitations, especially capturing complex material behaviors such as creep and plasticity before a thermo-mechanical fatigue model can be applied successfully with acceptable accuracy.

TREND

Wire bond, die attach, and heat sink attach have been the common failure mechanisms during power cycling or thermal cycling. Several packaging technologies are emerging to improve thermo-mechanical reliability for power electronic components.

New On-Chip Interconnects

Fuji Electric improved Al wire bond reliability by changing the re-crystallization temperature of the Al wire, limiting the growth of grains (Satheesh, 2014). Infineon replaced Al wire bonds with Cu wire bonds (Rusche, 2016). Because Cu has 'much higher yield and tensile strength combined with a lower CTE, the amount of plastic deformation for a given temperature cycle is significantly reduced compared to aluminum.' ROHM (ROHM Co. Ltd, 2024), Bosch (Durand, 2016), and Denso (Shimozuma, 2017) replaced Al wire bonds with soldered copper clips. Semikron (Semikron Ltd, 2011) used sintered foil as the on-chip interconnects. Hitachi (Hitachi, 2017) has replaced Al wire bonds with Cu wire bonds and a sintered buffer layer, which resulted in a 10× increase in power cycle life. All these new technologies have significantly improved the reliability of power electronic packaging and sometimes even shifted the weakest link to other locations.

Ag Sintering

Solder alloys are traditional die attach and heat sink attach materials for power modules. Solder joint fatigue is one of the most limiting factors for EV power electronic reliability. Typical solder creeps at room temperature due to its low melting point (½ T_m=~250 K). Thus, creep deformation is a significant failure mechanism causing power module failure. Solders are not suitable for application temperatures above 80% of their melting point (calculation is done in Kelvin). This means for a lead-free solder alloy with a melting point of 493 K (220°C), the application temperature needs to be below 394 K (121°C). Several studies (Schaal, 2018; Weber, 2016; Khazaka, 2014) have stated it is not desired to have the die attach operating temperature as high as 150°C for a very long time.

As EV power electronic systems move toward higher power density, the demand for joints that can survive higher operating temperature (>150°C) are increasing. Ag sintering is emerging as the better alternative (Ackerman, 2019). Ag sintering joints support higher operating temperatures greater than 250°C. TESLA, VW and BYD have transitioned from solder joints to Ag sintering joints (Wang, 2024). The list also includes many EV power module suppliers such as ROHM (ROHM Co. Ltd, 2024), Semikron Danfoss, Infineon, and FUJI (Horio, 2012).

The sintering process is a combination of temperature, pressure, sintering dwell time, heating rate, and sintering atmosphere. The silver particles of the sinter paste densify by means of diffusion processes. When the temperature increases above 230–250°C, together with an applied pressure of between 5 and 40 MPa, the silver paste density increases. The particles will bond together far below the melting point of Ag (962°C). This is a solid diffusion process, and Ag does not need to be melted. The pressure ensures that porous areas in the compound layer are reduced, which is particularly important for large-area semiconductors.

Ag particles used in the sintering process are micro- or nano-sized. The driving force for sintering is the tendency to minimize particle surface energy. Smaller particles have higher surface energy, which tends to lower the overall energy by merging those nano- or micro-sized particles into larger ones. The smaller the particle size, the higher the surface area-to-volume ratio, the higher the free energy available to facilitate sintering, and the easier it is to form the bonding. Micro-Ag can sinter at temperatures significantly lower than its melting point with the assistance of pressure. Nano-Ag particles can sinter at even lower temperatures due to its higher surface energy with little or no pressure. 'In the silver sintering process, heat is applied to a silver paste, several reactions occur simultaneously, including grain growth, pore growth, and densification, resulting in a stronger bond.' (Rogers Corporation, 2017).

Ag sintering requires metallization of the substrates or die to which it is bonded free of oxides. The thickness of the metallization does not need to be more than 1 μm since diffusion of silver won't exceed 25–75 nm. Noble metallization such as NiAu, NiPdAu or Ag are typical choices (Ackerman, 2019). The larger the bonding area, the more difficult it is to sinter. One manufacturer has reported sintering up to 350×270 mm (Koelink, 2017).

A power module with sintering technology used by Semikron Danfoss (Semikron Danfoss, 2009) was compared with a power module with a soldered joint between the chip and DBC substrate. Under power cycling conditions, soldered modules start to have failures after ~30,000 cycles due to bond wire lift-off, and significant increase in V_{ce} (collector to emitter voltage) increases significantly. While the sintered power module did not observe any degradation until after 90,000 cycles (>2× reliability), Daryanani (2023) has reported power module die attach using Ag sintering over soldering has also resulted in an improvement of 17 times in the active temperature cycling test (45°C to 175°C) on DBC substrates. Dutt (2022) found replacing soldering and wire bonding with a fully sintered module has delivered significant reliability gains in the form of a ten-fold improvement in power cycling. It also helped to double the current capability.

Semikron Danfoss has further developed Double Sided Sintering (DSS) technology in their power modules (Stockmeier, 2011). A sintering joint is applied both on the top side of the die to replace wire bonds and back side to replace die attach. In addition, DSS also replaces heat sink attach. It has provided 25% higher peak current due to a larger area joint at the front side of the die (see Figure 6.9).

Pressfit

Pressfit is another popular power electronic interconnect technology that serves as an electrical connection between power modules and PCBs or busbars. They can be either signal connections or power connections (Buschkühle, 2007; Semikron Danfoss, 2024; Wolfspeed, 2024; Vincotech, 2021). Pressfit is adopted by many leading power module original equipment manufacturers (OEMs) including Infineon, Semikron Danfoss, ROHM, and Wolfspeed. It eliminates the soldering process used to connect signal pins to the PCB.

During the assembly process, the pressfit pins are pressed into the board. The detailed process video is available at Wolfspeed (2024) . During pressing, plastic deformation occurs in the pins. As a result, the rework of the press may not be possible and can only be reworked via soldering. The pins are copper alloy with plating. The pin plating and PCB surface plating are specified by manufacturers. Common PCB platings, such as immersion tin and ENIG (Buschkühle, 2007) are recommended.

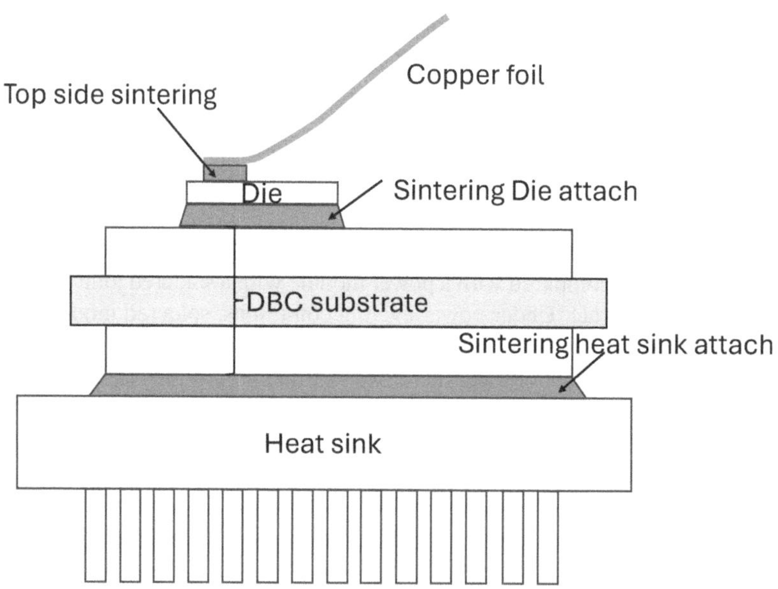

FIGURE 6.9 Illustration of Semikron Danfoss DSS technology.

The force required to press in and pull out the pins varies depending on the pin geometry. Due to plastic deformation, there is sufficient contact force to ensure low contact resistance for the connection. The contact resistance of the pressfit depends on the hardness and resistivity of the plating and the contact force. A minimum of 25 N/pin contact force has been reported, resulting in a very low contact resistance of 0.0055 mohm (Buschkühle, 2007). For a current of 20 A, the resulting voltage drop is 0.11 V and the power loss is 2.2 W. The power rating of the pressfit pin varies with pin design and temperature rise requirements; typically, each pin can handle 20–45 A of current.

Transfer Molded Package

Transfer molded package is another key technology that has emerged in recent years for EV power electronic modules. Bosch (Schoser, 2015) has found that transfer molded package tends to have better reliability because ‘the entire power stack is embedded in a mold compound that stabilizes the soldered stack and thus makes it extremely reliable against thermal cycles.’ Mitsubishi (Radke, 2015) used hard DP resin instead of soft silicone gel to cover wire bonds. The stress on the wire bonds due to CTE mismatch between wire bonds and die will be absorbed significantly by DP resin, unlike soft silicone gel. Also, the stress distribution is more even. Test results show that power modules with DP resin have seven times higher thermal cycling capability than conventional power modules with silicone gel.

BIBLIOGRAPHY

Ackerman, R. (2019, Jul 08). *The Next Step Sintering is the Key to Future Power Electronics.* Retrieved from EE Power: https://eepower.com/tech-insights/the-next-step-sintering-is-the-key-to-future-power-electronics/#

Alba, M. (2019, May 13). *Battery Bonding: Ultrasonic Wire Bonding vs Laser Welded Bonding.* Retrieved from Charged Electric Vehicles Magazine: https://chargedevs.com/features/battery-bonding-ultrasonic-wire-bonding-vs-laser-welded-bonding/

Barbagallo, C. E. (2017). Thermal fatigue life evaluation of SnAgCu solder joints in a multi-chip power module. *Journal of Physics,* 841

Bayerer, R. E. (2008). Model for power cycling lifetime of IGBT modules - various factors influencing lifetime. *International Conference on Integrated Power Electronics Systems* (pp. 1–6). Nuremberg, Germany: IEEE.

Borucki, L. E. (2018, Dec). *How Long Is Your System Going to Last? A Discrete IGBT Reliability Study Based on Power-Cycling Tests.* Retrieved from Infineon: https://www.infineon.com/dgdl/Infineon-IGBT_Reliability_Test_Bodos_Power_Systems-ART-v01_00-EN.pdf?fileId=5546d46269e1c019016a2ab43a7b2559

Buschkühle, M. E. (2007). *PressFIT for Solderless Assembly of Power Modules.* Retrieved from Power Electronics Europe, Issue 5: https://www.power-mag.com/pdf/feature_pdf/1222954420_PEE_Issue_5_2007_Power_Modules-PressFIT_for_Solderless_Assembly_of_Power_Modules.pdf

Daryanani, S. (2023, Oct 30). *Large-Area Sintering Advances Improve Power Module Performance.* Retrieved from Power Electronics News: https://www.powerelectronics-news.com/large-area-sintering-advances-improve-power-module-performance/

Durand, C. E. (2016). Solder fatigue failures in a new designed power module under power cycling. *Microelectronics Reliability*, *66*, 122–133.

Dutt, G. (2022, May 1). *Using Silver Sintering in Traction Inverter Assembly*. Retrieved from www.techbriefs.com: https://www.techbriefs.com/component/content/article/45721-using-silver-sintering-in-traction-inverter-assembly

Fuji Electric Co., Ltd. (n.d.). *Chapter 11 Reliability of Power Module*. Retrieved from https://www.fujielectric.com/products/semiconductor/model/igbt/application/box/doc/pdf/REH984e/REH984e_11.pdf

GopiReddy, L. E. (2014). Power cycle testing of power switches: A literature survey. *IEEE Transactions on Power Electronics*, *30*(5), 2465–2473. Retrieved from https://www.osti.gov/servlets/purl/1325461

Grams, A. (2015, Aug 1). *Numerical Modeling in Design for Reliability of Power Modules*. Retrieved from EE Power: https://eepower.com/technical-articles/numerical-modeling-in-design-for-reliability-of-power-modules/#

Guth, K. E. (2006). Improving the thermal reliability of large area solder joints in IGBT power modules. *4th International Conference on Integrated Power Systems*. Naples, Italy: VDE.

Hamm, R. E. (2015, Oct 1). *Maintaining Low Voiding Solder Die Attach for Power Die While Minimizing Die Tilt*. Retrieved from https://www.osti.gov/servlets/purl/1337975#:~:text=The%20best%20tooling%20combination%20for,combinations%20using%202%20weight%20rods

Harman. (1997). *Wire Bonding in Microelectronics: Materials, Processes, Reliability, and Yield*. New York, NY: McGraw-Hill Professional.

Hayashi, K. E. (2002). Improvement of fatigue life of solder joints by thickness control of solder with wire bump technique. *52nd IEEE ECTC*, 28–31 May, San Diego, CA.

Held, M. E. (1999). Fast power cycling testfor insulated gate bipolar transistor modules in traction application. *International Journal of Electronics*, 86(10).

Herrmann, T. E. (2007). Power cycling induced failure mechanisms in solder layers. *2007 European Conference on Power Electronics and Applications*, Aalborg, Denmark. IEEE.

Hitachi. (2017). *Hitachi IGBT for Automotive Vehicles*. Retrieved from https://www.mjc-elektrotechnik.de/tn-misc/download.php?index=1&folder=dl&id=6&type=pdf

Hitachi. (n.d.). *Hitachi IGBT for Automotive Vehicles*. Retrieved from https://www.electronicspecifier.com/news/analysis/goodbye-to-bond-wires-new-packaging-technology-replaces-wire-bonding

Horio, M. E. (2012). *Packaging Technologies for SiC Power Modules*. Retrieved from FUJI ELECTRIC REVIEW Vol. 58 No. 2: https://americas.fujielectric.com/files/FER-58-2-075-2012.pdf

Infineon. (2023, 05 05). *TO-247PLUS in Reflow-Solderable Package Product Description, Simulations and Enclosure Recommendations*. Retrieved from https://www.infineon.com/dgdl/Infineon-AN2023-06_TO-247PLUS_in_reflow-solderable_package-ApplicationNotes-v01_00-EN.pdf?fileId=8ac78c8c8900bb57018900e07ce302af

Inseto. (2024). *Aluminium Bonding Ribbon*. Retrieved from Aluminu: https://www.inseto.co.uk/consumables/bonding-wire-ribbon-precision-metal-stampings-solder-preforms-by-coining/bonding-wire-and-ribbons-by-coining/aluminium-bonding-ribbon/

Khan, Z. (2018, Dec. 7). *Revisiting ISO 13485 For Wire Bonding Quality and Reliability*. Retrieved from EEWeb: https://www.eeweb.com/revisiting-iso-13485-for-wire-bonding-quality-and-reliability/#:~:text=Now%2C%20with%20greater%20emphasis%20on,to%20further%20support%20ISO%2013485.

Khazaka, R. E. (2014). Review on joint shear strength of nano-silver paste and its long-term high temperature reliability. *Journal of Electronic Materials*, *43*(7), 2459–2466.

Koelink, M. (2017, Feb 27). *Ag-Sintering as an Enabler for Thermally Demanding Electronic and Semiconductor Applications.* Retrieved from EE Power: https://eepower.com/technical-articles/ag-sintering-as-an-enabler-for-thermally-demanding-electronic-and-semiconductor-applications/#

Lederer, M. E. (2022). Accelerated fatigue testing and lifetime modelling of solder interconnects in power electronics. *5th International Spring Seminar on Electronics Technology (ISSE)* (pp. 1–6). Vienna, Austria: IEEE.

Lee, W. E. (2000). *Solder Joint Fatigue Models: Review and Applicability to Chip Scale Packages.* Retrieved from https://www.researchgate.net/publication/242797412_Solder_joint_fatigue_models_Review_and_applicability_to_chip_scale_packages

Liu, J. E. (2009). *Reliability consideration of low-power grid-tied inverter for photovoltaic application.* Retrieved from Reliability consideration of low-power grid-tied inverter for photovoltaic application

Lutz, J. E. (2008). Power cycling induced failure mechanisms in the viewpoint of rough temperature environment. IEEE *5th International Conference on Integrated Power Systems (CIPS)*, Nuremberg, Germany, pp. 1–4.

Mainka, K. E. (2011). *Counting Methods for Lifetime Calculation of Power Modules.* Retrieved from Power Electronics Europe, Issue 6 2011: https://www.power-mag.com/pdf/feature_pdf/1319730265_Infineon0611_Layout_1.pdf

Miric, A. E. (2015). Inorganic substrates for power electronics applications. *Engineering, Materials Science.* https://www.scribd.com/document/674242777/Inorganic-Substrates-for-Power-Electronics-2015-3-AMiric-PDietrich-V8.

Mitsubishi Electric. (2019, Dec). *Power Module Reliability.* Retrieved from Mitsubishi Electric: https://www.mitsubishielectric.com/semiconductors/powerdevices/application_notes/powermodule_reliability_e.pdf

Morozumi, A. E. (2003). Technologies for reliable power modules in electric vehicles. *Fuji Electric Journal, 76*(10).

National Institute of Standards and Technology. (n.d.). *Sn-Ag-Cu Properties and Creep Data.* Retrieved from NIST Material Measurement Laboratory: https://www.metallurgy.nist.gov/solder/clech/Sn-Ag-Cu_Overview.htm

Nishiura, A. E. (2007). *Fuji Electric Review Vol. 53 No. 3.* Retrieved from IGBT Modules for Electric Hybrid Vehicles: https://americas.fujielectric.com/files/r53-3-2.pdf

Olsen, D. E. (1979). Properties of Die Bond Alloys relating to thermal fatigue. *IEEE Transactions on Components, Hybrids, and Manufacturing Technology, 2*, I–2.

Onsemi. (2023, Mar). *Soldering and Mounting Techniques Reference Manual.* Retrieved from SolderRM/D Rev14: https://www.onsemi.com/pub/Collateral/SOLDERRM-D.PDF

Radke, T. (2015, Aug 1). *More Power and Higher Reliability by 7th Generation IGBT Module with New SLC Technology.* Retrieved from EE Power: https://eepower.com/technical-articles/more-power-and-higher-reliability-by-7th-generation-igbt-module-with-new-slc-technology/

Ramli, M. E. (2022). Formation and growth of intermetallic compounds in lead-free solder joints: A review. *Materials (Basel)*, 15(4), 1451: https://doi.org/10.3390/ma15041451

Rogers Corp. (2024). *Rogers Corp: Advanced Electronics Solutions.* Retrieved from Curamik Ceramic Substrates Product Information & Data Sheet: https://www.rogerscorp.com/-/media/project/rogerscorp/documents/advanced-electronics-solutions/english/product-information/curamik-ceramic-substrates-product-information-and-data-sheet.pdf

Rogers Corporation. (2017, Jul 25). *Silver Sintering vs Soldering in High Power Density Power Electronics Circuits.* Retrieved from Rogers Corporation Advanced Electronics Solutions: https://rogerscorp.com/blog/2017/silver-sintering-vs-soldering-in-high-power-density-power-electronics-circuits

ROHM Co. Ltd. (2024, Jun 11). *ROHM's New TRCDRIVE pack with 2-in-1 SiC Molded Module: Significantly Reduces the Size of xEV Inverters*. Retrieved from ROHM Semiconductor: https://www.rohm.com/documents/11303/13026034/Introduction_NewRelease_TRCDRIVEpack_e.pdf/99157c86-82de-a19f-94ba-c7ae2e7af186?t=1718085675617

Rusche, W. E. (2016, July). *Lifetime Analysis of PrimePACKTM Modules with IGBT5 and .XT*. Retrieved from Infineon, www.Bodospower.com: https://www.infineon.com/dgdl/Infineon-Lifetime_Analysis_of_PrimePACK_Modules_with_IGBT5_and_XT-Bodo-Article-v01_00-EN.pdf?fileId=5546d4625607bd13015644d56a480c07

Satheesh, N. E. (2014, Jul 25). *Trends in Automotive Power Electronics*. Retrieved from Power Systems Design: https://www.powersystemsdesign.com/articles/trends-in-automotive-power-electronics/30/7569

Schaal, M. (2018). Silver sintering in power electronics: The state of the art in material characterization and reliability testing. *7th Electronic System-Integration Technology Conference (ESTC)* (pp. 1–18). Dresden, Germany: IEEE.

Schilling, O. E. (2012). Power cycling testing and FE modelling focussed on Al wire bond fatigue in high power IGBT modules. *Microelectronics Reliability*, *52*, 2347–2352.

Schoser, S. (2015, May 01). *Highly Reliable Transfer-Molded Power Modules*. Retrieved from EE Power: https://eepower.com/technical-articles/highly-reliable-transfer-molded-power-modules/#

Semikron Danfoss. (2009). *SiNTER Technology for Power Modules*. Retrieved from PCIM 2009: https://www.power-mag.com/pdf/feature_pdf/1242899514_sinter-technology.pdf

Semikron Danfoss. (2024). *Press-Fit Technology: Fast and Easy Assembly of Module and PCB*. Retrieved from Semikron Danfoss Innovation & Technology: https://www.semikron-danfoss.com/innovation-technology/packaging-technology/press-fit-technology.html

Semikron Ltd. (2011, May 17). *Semikron - New Packaging Technology Replaces Wire Bonding*. Retrieved from Electronic Specifier: https://www.electronicspecifier.com/news/analysis/goodbye-to-bond-wires-new-packaging-technology-replaces-wire-bonding

Shimozuma, A. E. (2017). Packaging technology of power module for automotive applications. *2017 International Conference on Electronics Packaging (ICEP)*. Yamagata, Japan: IEEE.

Siemens. (2025, Jan 7). *Field Lifetime Estimation of Power Modules using Active Power Cycling*. Retrieved from https://community.sw.siemens.com/s/article/Field-lifetime-estimation-of-power-modules-using-Active-Power-Cycling

Siemens Digital Industry Software. (2024). *Field Lifetime Estimation of Power Modules Using Active Power Cycling*. Retrieved from Siemens Digital Industries Software White Paper: https://www.plm.automation.siemens.com/media/global/pt/CE-35078%20Field%20Lifetime%20Estimation%20of%20Power%20Modules%20Using%20Active%20Power%20Cycling%20wp_tcm70-69422.pdf

Steinberg, D. S. (2000). *Vibration Analysis for Electronic Equipment*, 3rd edition. New York: Wiley.

Stockmeier, T. E. (2011). SKiN: Double side sintering technology for new packages. *2011 IEEE 23rd International Symposium on Power Semiconductor Devices and ICs*. San Diego, CA, USA: IEEE.

Syed, A. A. (2004). Accumulated creep strain and energy density based thermal fatigue life prediction models for SnAgCu solder joints. *2004 Proceedings. 54th Electronic Components and Technology Conference*. Las Vegas, NV, USA: IEEE.

Vincotech. (2021). *Vincotech's Press-fit Pin Application Note No: AN_2021-06_001_v06.* Retrieved from Vinotech: https://www.vincotech.com/fileadmin/user_upload/content_media/documents/pdf/support-documents/application-notes/Vincotech_AN_2021-06_001-v06_Press-fit.pdf

Wang, M. E. (2016). Effect of die size and die tilt on solder reliability under thermal cycling. *IEEE 66th Electronic Components and Technology Conference (ECTC).* Las Vegas, NV, USA: IEEE.

Wang, Y. (2024, Jun 21). *The Rise of Ag Sintering and Cu Sintering.* Retrieved from www.idttechex.com: https://www.idtechex.com/en/research-article/the-rise-of-ag-sintering-and-cu-sintering/31286

Weber, C. E. (2016). Combination of experimental and simulation methods for analysis of sintered ag joints for high temperature applications. *IEEE 66th Electronic Components and Technology Conference (ECTC).* Las Vegas, NV, USA: IEEE.

Wolfspeed. (2024, Apr 1). *Wolfspeed WolfPACK™ Power Module Press-In Process.* Retrieved from Wolfspeed: https://www.wolfspeed.com/knowledge-center/article/wolfspeed-wolfpack-power-module-press-in-process/

Yang, L. E. (2013, Mar). Physics-of-failure lifetime prediction models for wire bond interconnects in power electronic modules. *IEEE* Transactions on Device And Materials Reliability, *13*(1), 9–17. Retrieved from https://nottingham-repository.worktribe.com/prhttps://eepower.com/technical-articles/numerical-modeling-in-design-for-reliability-of-power-modules/#eview/1002793/Agyakwamodules.pdf

Yang, S. E. (2011). An industry-based survey of reliability in power electronic converters. *IEEE Transactions on Industry Applications, 47*(3), 1441–1451. Retrieved from https://www.researchgate.net/publication/224224489_An_Industry-Based_Survey_of_Reliability_in_Power_Electronic_Converters

Zeng, G. E. (2018, June). *First -Results of Development of a Lifetime Model for Transfer Molded Discrete Power Devices.* Retrieved from PCIM Europe 2018: https://www.infineon.com/dgdl/Infineon-PCIM_2018_First_results_of_development_of_a_lifetime_model_for_transfer_molded_discrete_power_devices-Article-v01_00-EN.pdf?fileId=5546d4626f229553016f89a3cffd14b9

Zheng, Y. (2005). *Effect of Surface Finishes and Intermetallics on the Reliability of SnAgCu Interconnects.* College Park, MD: University of Maryland.

7 EV Power Electronic System Vibration

INTRODUCTION

Steinberg (2000) pointed out, 'Most electronic failures are mechanical in nature, many of these mechanical failures occur in the component lead wires and solder joints. Extensive military testing experience over a period of many years has shown...about 20% are due to some form of vibration and shock.' The EV electronic drive system is under vibration excitation from the road and from the drive system, including motor and transmission. Vibration failure occurs when the stress under vibration exceeds the vibration strength of the component. Many electronic components and connections are susceptible to vibration failure. Normally, the components on the PCB and the connections are the most vulnerable.

Vibration design principles are applicable to both traditional ICEs and EV/HEV powertrains. As the electric drive system is moving toward higher power density and higher integration, the inverter is generally integrated with the motor and transmission. This leads to new vibration challenges, such as more intensive HF excitations, while the overall structure is becoming lighter.

FACTORS AFFECTING VIBRATION RELIABILITY

There are three aspects of vibration. The first is the excitation, the source of the vibration. The second is how the vibration is transmitted. The focal point of vibration transmission is resonance when the response is amplified, which could result in excessive deformation or even failure. The last aspect is the vibration resistance of components, which is the capability of the components to withstand vibration deformations without failure.

EXCITATIONS

Excitations are the sources of vibrations. For automotive applications, one main source of vibration excitation is from the road during driving. Rougher roads will introduce higher acceleration and more vibration stress. Typically, the vibration related to road noise is of low frequency (<100 Hz) and predominantly random in nature.

In EV/HEV applications, direct mounting of traction inverters on the drive unit is becoming a mainstream choice. The list is long: Chevrolet Volt (2016), Tesla, Audi Etron, Hyundai E-GMP, Nissan Leaf, VW ID4, BMW i3, and Lucid. This

 DOI: 10.1201/9781003569961-7

minimizes the connections, saves space, reduces weight, and enables better performance. However, vibration excitations from the driving unit are becoming a major vibration source for inverters. Those excitations are produced by rotational components either from the motor or from the transmission. Thus, they are sine-on-random in nature. Typically, the excitation frequencies are higher than 100 Hz (see Figure 7.1).

In EVs, there are no engine excitations such as piston unbalance in either four- or six-cylinder engines. Instead, the vehicle excitations come from the drive unit. The vibrations which the inverter is subjected to can differ significantly depending on where it is mounted, whether the vehicle has an engine,electric motor, or transmission, the characteristics of the mounting structure, and suspension performance.

Traction motors are a major vibration source. First, the imbalance of the rotor or shaft rotating introduces a centrifugal force due to uneven mass distribution, leading to vibrations. Secondly, a misaligned shaft can generate significant vibrations. Bearing problems could also produce excessive vibrations. For example, bearings will be worn over time and can cause rotor vibration. Loose, improperly balanced motor mounting could also introduce significant vibrations. Another source is from the electrical system. AC harmonics from the inverter produce the torque ripple in the motors, which directly translates to mechanical vibration. Variation of the air gap between the rotor and the stator can also introduce vibrations. The transmission system is another well-known source of vibrations.

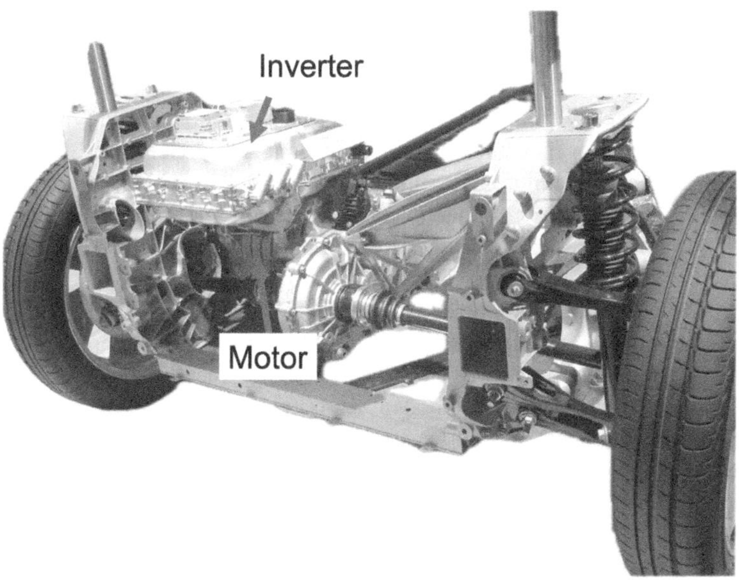

FIGURE 7.1 In the BMW i3 Drive Unit, the inverter is directly mounted on the traction motors (Rudolf Simon, licensed under the Creative Commons Attribution-Share Alike 3.0 Unported license from https://commons.wikimedia.org/wiki/File:BMW-i3-Electric_Drive_unit2.JPG).

Resonance

Vibration is largely about resonance. When the excitation frequency matches the natural frequency of the system, the response is amplified, which introduces excess stress and strain and eventually damages the components. In power electronics components, a lot of vibration failures are related to PCB resonance. PCBs are flexible and easily resonate with low frequency excitations. In the meantime, many electronic components on PCBs are vulnerable to vibration failures.

Typically, PCBs have fundamental frequencies between 100 and 300 Hz. A significant amount of PCB resonance occurs when power electronic components are mounted on the drive unit with excitation frequencies greater than 100 Hz.

If the frequency of the excitation is around the resonance of the components, especially at the fundamental resonant mode, the displacement of the component will be greatly amplified and create excessivedeformation. As Steinberg (2000) has pointed out, 'The fundamental resonant mode of the system often has the greatest displacement amplitude and usually the greatest stresses. Most of the damage in electronic assemblies occurs at the fundamental natural frequency of the PCB.'

Resonance between the PCB and connecting components, especially its supporting structure, is also critical. The resonance introduces amplified acceleration and excessive damage. If the natural frequency of the supporting structure is close to the natural frequency of the PCB, the response can amplify the PCB vibrations, resulting in positive feedback and producing significantly higher accelerations on the PCB.

Components Vibration Resistance

Compared to mechanical components, electrical components are normally not designed for vibration robustness. Connections and components on PCBs are very vulnerable to vibration failures. Component characteristics such as weight, stiffness, and strength of the joint or connections affect the vibration reliability. Some components and joints are less resistant to vibration stress. Capacitors are among the components most susceptible to vibration failures.

Interconnect vibration fatigue is one of the most typical vibration failure mechanisms. Interconnects include solder joints, leads connecting components to PCBs, and gate connections from power modules to PCBs. Typically, interconnects (solder joints) or connections such as component leads connect components to the PCB. In the design, interconnects are the least stiff portion. They are strained the most to accommodate and absorb the differences since two parts respond differently under vibration (see Figure 7.2). Damage accumulates at the interconnects over time, eventually leading to vibration fatigue failure. On the other hand, if interconnects are too stiff to absorb the deformation, the components, such as brittle ceramic capacitors, have to withstand the deformation. Vibration failure may occur to the ceramic components instead. Heavy components such as transformers and capacitors can cause higher stress to the interconnects and components themselves under the same excitation (see Figure 7.2).

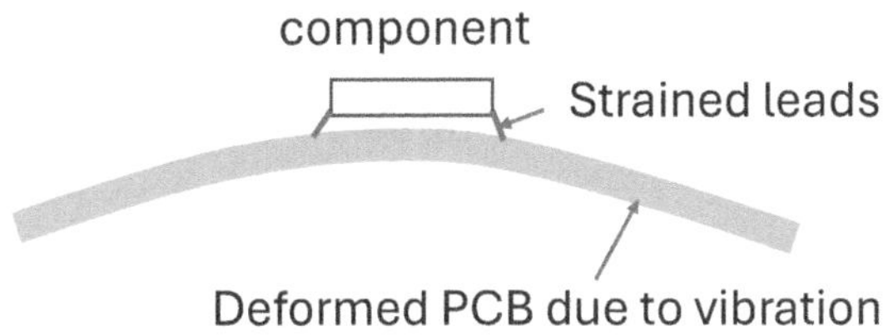

FIGURE 7.2 Component leads were strained due to relative motion between the PCB and component during vibration.

VIBRATION CHARACTERIZATION

To ensure components can survive their lifetime (10 years/100,000 miles), accelerated vibration tests are performed extensively with electrodynamic shaker or hydraulic shaker rigs. The test is expected to produce the same failure as that experienced in real applications within a fraction of the actual time (tens of hours). A test signal from either standards or real applications in the form of acceleration power spectral density (PSD) can be used as input to the accelerated vibration test.

PSD

PSD is a parameter to quantify random vibration (Siemens, 2020). It is a measure of vibration signal intensity in terms of g^2/Hz over frequency. High PSD at a certain frequency range means a large amount of vibration energy concentrated at this frequency range, often linked to resonances or excitation source frequencies.

ISO Standards

Vibration profiles from ISO vibration standards such as ISO 16750-3 and ISO 19453-3 are widely used in the automotive industry for vibration durability testing (ISO, 2012).

Vibration affecting electronic components differs significantly depending on mounting locations, for example, whether it is mounted on the engine, chassis, and motors, as well as suspension system. As shown in the Table 7.1, vibration profiles vary with types of vehicles (passenger car or commercial). They also vary with excitation sources, whether they are road or EV drive systems.

In these standards, there are different profiles corresponding to different vehicle and mounting conditions. Each profile typically considers three sub-profiles:

- Sinusoidal excitation due to engine, motor, and gearbox
- Random vibration due to engine and gearbox non-sinusoidal vibrations
- Random vibrations due to road excitations

Sometimes, those sub-profiles are combined. Motors, engines, and gearboxes have rotating components like rotors, crankshafts, and gears. Component rotation generates sinusoidal vibration when imbalance or misalignment occurs.

TABLE 7.1
List of Several Vibration Sets in ISO 16750-3

	Sinusoidal Vibration	Random Vibration Due to Vehicle Excitation	Random Vibration Due to Road Excitation
Passenger Car, Component on Engine	From the unbalanced mass forces	Yes. Due to all other vibration-schemes of an engine	Yes
Passenger Car, Component on Gearbox	Due to unbalanced mass forces of engine	From the friction of the gear wheels and other schemes, and low frequency road excitations.	
Passenger Car, Component on Flexible Plenum Chamber	Due to pulsation of the intake air	N/A	N/A
Passenger Car, Sprung Mass (Vehicle Body)	N/A	N/A	Yes
Passenger Car, Unsprung Masses	N/A	N/A	Yes
Commercial Vehicle, Engine or Gearbox	From unbalanced engine mass forces	From random noise due to all other engine sources	

N/A: not applicable here

Most of the EV vibration excitations are random in nature, regardless of road excitations or drive unit excitations. The excitations occur across a wide range of frequencies. The road excitations are mostly low frequencies (<100 Hz). On the other hand, drive unit excitations (engines and gearbox) mostly occur at HF range (>100 Hz). Therefore, vibration excitation produced by high-speed driving on flat roads is mostly determined by the drive unit and is characterized by high PSD at frequencies higher than 100 Hz. Low-speed driving on rough roads produces vibrations mostly due to the rough road, indicated by high PSD at frequencies lower than 100 Hz (see Figure 7.3).

ISO 19453-3

ISO 19453-3 (ISO, 2018) is an ISO standard profile specifically for electrical and electric equipment for the drive system of electric propulsion vehicles. (International Standard, 2018) There are several profiles depending on the applications:

The first profile is applicable to passenger cars with powertrain (combustion engine, gearbox).

Three types of sub-profiles are included:

a. Sinusoidal vibration due to unbalanced mass forces in the cylinders
b. Random vibration due to all other vibration schemes of an engine
c. Random vibration due to the influence of rough road (<100 Hz)

The second profile is applicable to passenger cars, and vibration is due to sprung mass (vehicle body). It is mainly due to road excitations, and high PSD occurs at frequencies less than 100 Hz.

The third profile is applicable to EVs equipped with an electric motor.

The motor vibration is mainly due to rough road driving. The motor only needs to consider excitation from the rough road and does not need to consider the vibration from the motor itself. Therefore, the vibration profile of motors is different from the profile of the inverters mounted on the motor.

ISO 16750-3

ISO 16750-3 vibration describes the vibration loads that affect electric and electronic systems and components with respect to their mounting direction and vehicle types. Some profiles are listed in Table 7.1.

The profiles indicate that for random vibration, the difference in terms of vibration intensity (PSD) for specific frequencies among these different profiles can be as huge as 100 times difference in the PSD value for a certain frequency.

How to pick the right ISO vibration profile depends on the design objective. An inverter OEM will likely select a profile that will match the applications. For example, the Hitachi inverter used two standard profiles based on ISO 16750-3 and JIS-D1601 (Nakatsu, 2014). One for chassis mounted, which includes high PSD (>1 $(m/s^2)^2/Hz$) at frequencies <100 Hz. For engine mounted components, the profile is ISO 16750-3 P1 (passenger car, engine). It has high PSD (20 $(m/s^2)^2/Hz$) under HF(≥500 Hz) and 10 $(m/s^2)^2/Hz$ under low frequency (≤100 Hz). Vibration under engine mounted components (18.2 Grms) is more severe than chassis mounted (2.8 Grms).

It is important to understand the applications and choose an appropriate profile. Most ISO standards are expected to be more severe than the actual vibration the components experience and cannot represent the real vibration profile. It would be ideal to obtain the vibration profile for each component by field testing. However, oftentimes the actual profile is not available during the component development stage. Developing inverters with vibration robustness needs to be completed long before new vehicles can be driven on the road and vibration excitation can be measured. Therefore, it is a good start to use the ISO standard before the real profile becomes available.

Mission Profile

Vibration excitations to the inverter vary from one driving condition to another. It would be ideal to determine the real excitations from every driving condition by testing, but it is not realistic. Mission profiling and test synthesis is a process to design a set of simple tests including a few representative driving conditions that can produce the same failure as the real environment, but in a fraction of the real time.

Vibration profiles can be fairly easily measured at the locations where these components are mounted to the main frame. The process to determine the profile is described here (Halfpenny, n.d.):

Step 1 is to determine representative drive events and corresponding occurrences depending on the application.

The mission profile needs to include several representative events experienced by the vehicle during various real-life situations. For example, it is critical to consider typical powertrain events and occurrences for the inverter placed on the drive unit. Table 7.2 is a list of some representative events.

Those events represent what inverters experience from the motor and gearbox during driving. For example, WOT refers to acceleration or towing using the maximum torque available. US06 is an aggressive driving event that represents high acceleration and high-speed driving. UDDS is a driving test that represents city driving with frequent stops. Both 20% grade and 30% grade driving involve driving on 20% grade or 30% grade respectively with WOT.

For other components such as an OBC, when it is mounted on the chassis, the road excitation may be more critical. Standard vibration test events for those components will mainly consider the impact from road excitations. The events may include different rough roads.

Step 2 is to place accelerometers in critical locations on the vehicle and collect responses (acceleration changes over time) under the driving events specified in Step 1.

TABLE 7.2
Examples of Various EV Powertrain Events

Powertrain Events	Test Speed	Estimated # of Cycles for 100 K Miles
WOT to 60 mph, maximum torque	0–60 mph	8000
Partial open throttle (POT) to 80 mph, 70–80% torque		3000
Passing maneuver WOT	50–80 mph	7000
US06 load	High speed	10,000
UDDS	Start/stop	20,000
Sustained V_{max}	Max speed	8000
Regen braking	From 80 to 0 mph	5000
20% Grade	WOT	1000
30% Grade	WOT	800
Reverse 20% grade		800

The location of the accelerometer placement is critical for the test to determine representative vibration behavior. Accelerometers are typically mounted on the component mounting feet to the drive unit or vehicle chassis. These are the locations where excitations are transferred through in actual application and vibration testing with a shaker table.

Neither drive units nor components are rigid bodies like shaker tables. Therefore, different mounting locations will have different vibration profiles. It is important to evaluate the impact of different locations and determine the most representative one. The accelerometer measures both excitations from the drive unit or road and the resultant response at the location. Typically, at mounting locations, responses are expected to be dominated by excitations. Therefore, the measurement can be used as excitations.

Step 3 is to use vibration analysis software such as nCode. It can convert test data from the time domain to the frequency spectrum based on a certain shock response spectrum. Then it can combine all the events (mission profile) and is equivalent to an accelerated test profile under a specified test duration based on the S-N fatigue model. The S-N fatigue model for vibration is an inverse relationship between maximum vibration stress amplitude (ΔG_{rms}) under repeated loading and the number of cycles to failure (N_f) on a logarithmic scale. It is normally established based on test data for a given material.

Step 4 is the execution of applying the vibration profile obtained to the parts mounted on the shaker table. Typically, the test is performed in three axes separately for the specified test duration. Parts are qualified in terms of vibration durability if they can pass the test without failure.

HOW TO DESIGN VIBRATION PROOF COMPONENTS

The principle to reduce vibration failure in electronic systems is to reduce the stress/strain in the components due to vibration and to increase the resistance of the component against vibration failure.

Increase the Fundamental Frequency of the PCB

In GMW 3172 (General Motors Corp, 2008), it is recommended that PCBs mounted on the system should have a resonant frequency greater than 150 Hz. Increasing the fundamental frequency of PCBs above 150 Hz can avoid resonance from low-frequency excitations, improving the vibration reliability of components on the boards (Habtour, 2013). This is particularly helpful for components mostly subjected to road excitations, which are generally less than 100 Hz.

Generally, components with low fundamental frequency experience larger dynamic displacement under excitation and therefore are more likely to fail. Designing components with higher fundamental frequencies is therefore preferred. This is just a rough design guideline. In actual application, some structures may be subjected to much higher excitation. Some components may be less sensitive to vibration. Therefore, the design guideline will need to be adjusted accordingly. In the end, testing is needed to validate the design.

Frequency Separation

Resonance frequencies of the PCB supporting structure are preferred to be well separated from the PCB fundamental frequency by a factor of two (Steinberg, 2000). The goal is to minimize resonance between the PCB and its supporting structure. However, this may become a challenge, especially for small power electronic systems where there is not much space left to separate the resonant frequencies.

As explained earlier, increasing PCB fundamental frequency is preferred to reduce displacement and reduce vibration risk. The fundamental frequency of the PCB is normally lower than the fundamental frequency of the supporting structure in power electronic systems. Increasing the resonant frequency of the PCB brings it closer to the resonance frequency of its supporting structure, which may adversely lead to a higher possibility of resonance. Therefore, there is always a trade-off between increasing the fundamental frequency of the PCB and the separation of fundamental frequency between the PCB and the supporting structure (see Figure 7.3).

Resonance frequency of a mechanical system can be described with the equation below.

$$f_0 = \frac{1}{2\pi}\sqrt{\frac{k}{m}}$$

f_0: resonance frequency
k: stiffness
m: mass

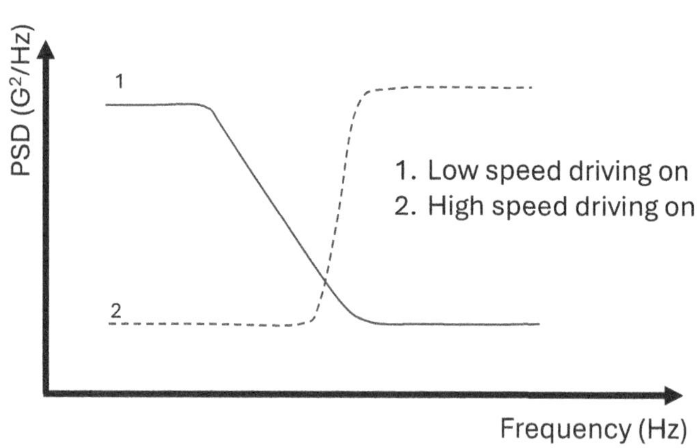

FIGURE 7.3 Typical vibration profiles include a sine-on-random covering engine vibration in high-speed driving and random vibration in low-speed driving conditions due to road excitations.

There are two options to adjust the resonance frequency. One is to change the mass. The other is to adjust the stiffness of the structure. PCBs are composite materials with layers of FR4 and thin copper. Their weight is dictated by the electrical layout. There is less flexibility to change the weight. Modifying the stiffness is a more practical solution. Critical factors determining stiffness are elastic modulus and constraints. The elastic modulus of PCBs depends on the number of copper layers and the area of copper in each layer. The elastic modulus is mostly dictated by the electrical layout. Therefore, it is difficult to adjust. Instead, adjusting the constraints of PCBs is more feasible to achieve the desired PCB vibration response. In automotive applications, a very effective practice to constrain PCBs is using constrained PCBs with fasteners.

A case study is explained to illustrate the principle. Figure 7.4 shows a PCB (area 100 mm×150 mm, thickness 1.6 mm, E=20 GPa, density: 0.0027 g/mm^3). With four fasteners on the four corners (red), the fundamental frequency of the PCB is ~82 Hz. Adding one fastener in the center significantly increases the fundamental frequency to 200 Hz. The frequency is estimated based on equations provided on page 145 (Steinberg, 2000).

One limitation of applying fasteners is that they take up too much space. For example, an M4 fastener would occupy an area of 8×8 mm. Therefore, although more fasteners on a PCB are typically preferred to achieve a higher fundamental frequency, the reality is that the spacing available for fasteners is always limited as inverters are becoming smaller and smaller. Thus, it is always desirable to place fasteners in the most effective locations without sacrificing electrical performance.

As package size becomes smaller, some modules use signal pins or pressfit pins to provide some level of constraint. This option needs to be implemented with care because signal pins are not expected to sustain too much stress. Certain stress release methods may be needed depending on the type of pins to minimize stress due to either assembly or vibration.

Adjusting resonance frequencies of the supporting structures is also important. Bigger and thicker housings are usually stiffer. Adding more fasteners at appropriate locations is helpful as well. As supporting structures become smaller, their stiffness increases and mass also reduces, which may lead to higher resonant frequencies for both the supporting structures and the PCB.

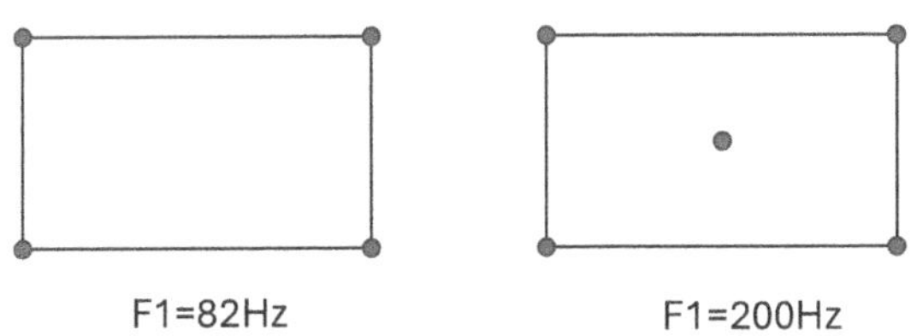

FIGURE 7.4 The fundamental frequency of the four corner fastened PCB increased from 82 to 200Hz by adding one fastener in the middle.

High vibration-resistant components

Another solution to increase system vibration durability is to improve individual components' capability to withstand vibration. First, a standard of acceptability is needed to quantify components' vibration resistance (EEEPower, 2020). This is difficult for EV applications because vibration is normally random in nature, and the source may come from the drive system and from road surfaces. This random vibration is difficult to capture at component level. Instead, sinusoidal excitation, sweeping over a certain frequency range, is used to quantify a component's vibration resistance. IEC 60068-2-6 sinusoidal vibration testing (2008) is a standard that can serve the purpose of evaluating if certain components have a certain level of vibration capability. It specifies the sinusoidal vibration sweep over a certain frequency range for a given period. The components that can still operate after going through the test are certified to meet the standard (EEEPower, 2020). For example, for general-purpose land-based transport applications, the standard specifies a sinusoidal vibration over 10–55 Hz and 1.5 mm peak-to-peak displacement, with ten sweep cycles in each axis (2 hours on each axis). The acceleration amplitude can be 5 or 10 g. OEMs can design components to meet or exceed this standard for better vibration performance. This is commonly used for capacitors, which rank among the most vulnerable electronic components to vibration damage.

One example is Panasonic's vibration-proof capacitors. It can withstand 30 g instead of 10 g acceleration and a peak-to-peak displacement of 5 mm with sinusoidal vibration excitations sweeping from 5 to 2000 Hz on three axes (2 hours on each axis) without failure (EEEPower, 2020). The design incorporates special features to improve its vibration performance. It includes 'thicker internal connections, high walls to the plastic baseplate, and more supportive terminal contacts.'

Another example is multilayer ceramic capacitors (MLCC). MLCCs are susceptible to flex cracking. When they are mounted on the PCB, bending of the PCB due to vibration will transfer the load to the ceramic body, which is stiff and susceptible to cracking over time. One solution by KEMET and AVX is to provide MLCCs with flexible termination (Hammerschmidt, 2020) .

Providing components more resistant to vibration failure is always appealing to automotive applications. If components are vulnerable, locating those components in the area that has less displacement under vibration is an alternative option (Nakatsu, 2014).

Damping

Fasteners, ribs, thickness, and weight adjustment are typical options to ensure the resonance frequency of the systems is well separated to minimize vibration failures. However, as power electronic packaging moves toward higher power density and more compact designs, it is becoming more challenging to make the above design modifications. Other options to mitigate vibration damage are needed. One example is vibration damping electrical connectors used in the 2016–2022 Toyota Prius inverter (Kelly, 2017). These connectors are connections between gate drive boards and power inverter module signal pins. They are designed to reduce vibration-induced displacement.

Other methods

The EV Motor is controlled by PWM switching signals from the traction inverter. The AC signals are not perfect and have many harmonics. The low-frequency harmonics can introduce torque ripple and vibrations. Optimization of the motor control algorithm can minimize torque ripple and vibrations. Toyota (Ito, 2005) developed motor control methods to reduce vibration for hybrid vehicles. One method is to prevent the engine torque ripple from being transmitted to the wheels. The other is to reduce the torsional vibration of the drive shaft. This is realized by providing compensation motor torque through motor control software.

METHODOLOGY FOR VIBRATION DESIGN OPTIMIZATION/VALIDATION

Computer-aided engineering (CAE) such as modal analysis has been proven to be an effective method to optimize PCB resonance response. In addition, CAE can be used to estimate the vibration life of the component interconnects. Modal analysis is an effective method to detect structural weaknesses that may lead to vibration fatigue, especially on the PCB. It can be used to determine the frequency characteristics of resonance and stress on the system (Nakatsu, 2014). But CAE alone cannot validate the design. CAE needs to be integrated with testing to enable efficient and effective design validation.

BIBLIOGRAPHY

EEEPower. (2020, Sep 10). *Vibration-Proof Capacitor Technology.* Retrieved from https://eepower.com/technical-articles/vibration-proof-capacitor-technology/

General Motors Corp. (2008, Aug). *GMW3172 GM Worldwide Engineering Standards.* Retrieved from https://www.scribd.com/document/305450584/GMW3172-GM-Worldwide-Engineering-Standards

Habtour, E. (2013). Physics of Failure for Portable Electronic Devices in Military Applications. *2013 Proceedings Annual Reliability and Maintainability Symposium (RAMS).* Orlando, FL, USA: IEEE. Retrieved from https://ris.utwente.nl/ws/portalfiles/portal/113172707/Habtour2013physics.pdf

Halfpenny, A. (n.d.). *Accelerated Vibration Testing Based on Fatigue Damage Spectra.* Retrieved from http://www.vibrationdata.com/tutorials_alt/fatigue_damage_spectra.pdf

Hammerschmidt, C. (2020, Jul 20). *How Antivibration Capacitors Transform the Reliability of ECUs.* Retrieved from https://www.eenewseurope.com/en/how-antivibration-capacitors-transform-the-reliability-of-ecus/

International Standard. (2018). *ISO 19453-3:2018(E) Road Vehicles — Environmental Conditions and Testing for Electrical and Electronic Equipment for Drive System of Electric Propulsion Vehicles Part 3 Mechanical Loads.* Retrieved from https://www.sis.se/api/document/preview/80003195/

ISO. (2012, Dec 15). *ISO 16750-3 Road Vehicles - Environmental Conditions and Testing for Electrical and Electronic Equipments.* Retrieved from https://www.sis.se/api/document/preview/915636/

ISO. (2018). *ISO 19453-3: 2018 (EN) Road Vehicles — Environmental Conditions and Testing for Electrical and Electronic Equipment for Drive System of Electric Propulsion Vehicles.*

Ito, Y. e. (2005, Mar. 28). *https://www.tytlabs.co.jp/en/english/review/rev402epdf/e402_037ito.pdf.* Retrieved from R&D Review of Toyota CRDL Vol. 40 No 2: https://www.tytlabs.co.jp/en/english/review/rev402epdf/e402_037ito.pdf

Kelly, J. (2017). *ATTC 4530 Hybrid and Electric Vehicle Systems Week 7 - 4th Gen. Prius Inverter with Converter.* Retrieved from https://www.youtube.com/watch?v=qcW08Mh-eIc

Nakatsu, K. E. (2014). *Next-Generation Inverter Technology for Environmentally Conscious Vehicles.* Retrieved from https://www.hitachihyoron.com/rev/pdf/2012/r2012_06_107.pdf

Siemens. (2020). *What Is a Power Spectral Density (PSD)? How Is It Different Than an Autopower?* Retrieved from https://community.sw.siemens.com/s/article/what-is-a-power-spectral-density-psd

Steinberg, D. (2000). *Vibration Analysis for Electronic Equipment* (3rd ed.). Wiley-Interscience.

8 EMI in Electric Drive Systems

ABSTRACT

EMI is the disturbance from electromagnetic emissions that affects the normal operation of an electronic device or system. EVs, especially electric drive systems, are vulnerable to EMI. EMI can interfere with the normal operation of electronic components and systems within the vehicle. This interference can lead to malfunctions, degraded performance, or even complete system failure if not properly addressed. EV electric drive systems are especially concerning since they use HV systems (400–800V motors and inverters) that switch large currents (hundreds of amperes) rapidly with PWM switching at several kHz. The chapter discusses EMI in EV electric drive systems. Critical factors, such as power device switching, parasitic components, and cables, will be addressed. A typical mitigation method is briefly reviewed at the end.

INTRODUCTION

Main sources of EMI in electric drive systems are related to the rapid switching of power devices and parasitic components (Taranovich, 2023). Power devices used in several critical EV components operate at high switching frequencies and high power. Those components are

- Converters (OBC and DC-DC)
- Inverters
- Motors

Fast switching of those HV power components generates strong electromagnetic fields and becomes a major source of EMI. Studies (Moon, 2015) have found that PWM, the commonly used EV power electronic switching technique, can lead to serious EMI issues:

- Ground current escaping through stray (parasitic) capacitors
- Conducting and radiating EMI
- Inducing bearing current and shaft voltage

The higher the power of these components, the stronger the magnetic fields can potentially become. The higher the switching ramp rate, the more severe EMI can potentially occur. The EV power rating for these components ranges from a few to

DOI: 10.1201/9781003569961-8

hundreds of kilowatts, while switching frequencies range from several to tens of kHz. All those can introduce powerful magnetic fields.

EMI can also affect the source (component) itself, resulting in poor function and distortion of the signal. Electromagnetic emissions can be conducted via PCB traces, wires, and cables and then radiated away. This results in EMI, introducing unwanted noise or interferences to other devices and causing them to operate poorly, to malfunction, or to stop working completely. For example, EMI produced by HV components can interfere with LV communication systems in vehicles and disrupt the proper functioning of those systems.

EMI TYPES

There are two major types of EMI (see Figure 8.1) depending on the propagation path. One is the conducted EMI transmitted along the circuitry. It is the near-field coupling within the components and the circuitries. The other one is radiated EMI via cables or PCB traces. The cables and PCB traces in the vehicle serve as an antenna to transmit and receive EMI.

CONDUCTED EMI

Conducted EMI is the EMI noise coupled through electrical lines, either power lines or signal lines. It includes cables and power connections, along with parasitic impedances and ground connections. It may affect the performance of other connected devices, causing devices to reset or lose power if the circuit is overloaded. Conducted EMI typically ranges from 1 kHz to 30 MHz, but it can reach up to 108 MHz (Ramadass, 2023a). Conducted emissions are either conducted directly into the circuits, or capacitively or magnetically coupled to a circuit.

Two modes of conducted EMI are CM and DM. As Figure 8.2 indicates, CM noise current flows in power lines in the same direction on two or more conductors with

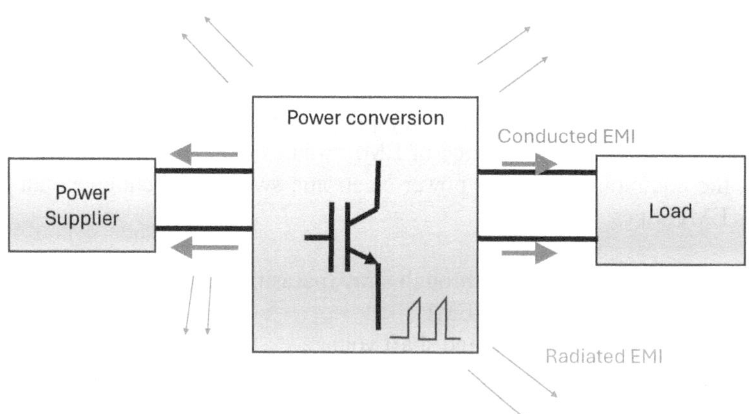

FIGURE 8.1 Illustration of conducted EMI and radiated EMI for power conversion systems.

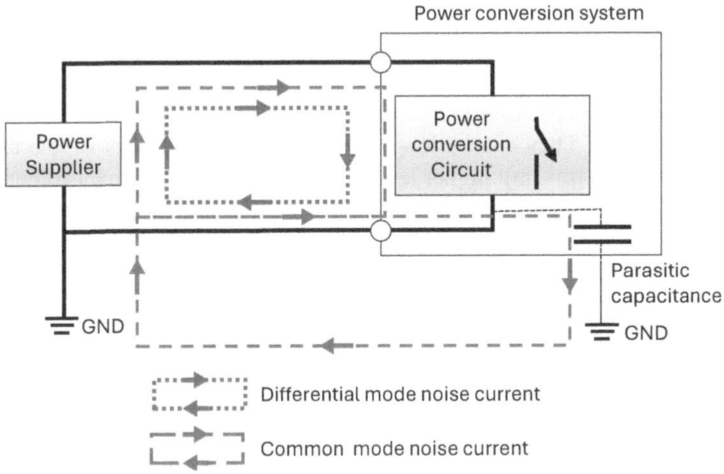

FIGURE 8.2 CM and DM noise current.

respect to a common reference point (e.g., ground). DM noise current flows in power lines in opposite directions on two conductors (e.g., the two wires of a twisted pair).

Common Mode EMI

In power electronics, CM noise is normally caused by a capacitive coupling of the switching nodes into the ground (e.g., vehicle chassis parasitic capacitance) and appears with the same phase and amplitude at both lines relative to the common reference point (see Figure 8.2). Usually, the common reference point is the ground. CM EMI is often associated with ground loops for EV applications. CM noise is typically driven by high dv/dt switching transients. The switching transients may couple through parasitic capacitance from a power electronic package heat sink to the ground, creating a CM current loop.

Differential Mode EMI

DM current noise (see Figure 8.2) is a current that appears on two lines of a closed loop, and the flow is in opposite directions. DM EMI is usually associated with the normal operation of electronic circuits. The noise is typically driven by high di/dt switching transients. One typical scenario happens to an inverter. First, high di/dt is generated by power switches. Then it couples with the parasitic inductance associated with the commutation loop between power switches and the DC capacitor. This introduces voltage spikes and creates a DM current noise within the loop.

Radiated EMI

Radiated EMI gets coupled through the air as radio waves due to electric fields (capacitive coupling) or magnetic fields (magnetic coupling). Radiated EMI covers frequencies from 30 MHz to 1 GHz, higher than conducted EMI (Ramadass, 2023a). It is located in the far field which is around two or more wavelengths distance from the source.

Circuit conductors (such as PCB traces, busbars, and cables) can serve as antennas to radiate EMI. They can be loop antennas, slot antennas, or dipole antennas. These antennas can transmit higher levels of EMI at certain resonant frequencies.

Based on dipole antenna theory, the electric filed radiation strength E created by CM current can be estimated for an electrically short monopole antenna (Herrick, 2009; ROHM semiconductors, 2018b):

$$E = \frac{4\pi \times 10^{-7} \left(fIl \right) sin\theta}{r} \tag{8.1}$$

E: uV/meter
f is the CM emission frequency
l is the length of the antenna (cm)
I is the CM noise current (amps)
r is distance from the antenna to the measurement point

- θ is the angle from the axis of the antenna, with maximum radiation at θ =90°

Based on loop antenna theory, the electric filed radiation strength E created by DM current can be estimated for a loop area A (Herrick, 2009; ROHM semiconductors, 2018b).

$$E = \frac{1.316 \times 10^{-14} \left(f^2 AI \right) sin\theta}{r} \tag{8.2}$$

f is the DM emission frequency
A is the DM current loop area
I is the DM noise current
r is distance from the loop to the measurement point
θ is the angle from the axis perpendicular to the loop plane

This is derived from Faraday's law. The magnetic field generated by a loop is directly proportional to the area of the loop and rate of current change.

Equation 8.1 and Equation 8.2 indicate how the radiated electromagnetic field changes with geometry, loop current magnitude, and frequency for loop antennas and dipole antennas. Based on the equations, radiation due to CM is typically far greater than the radiation due to DM for the same noise current levels (ROHM semiconductors, 2018a).

ELECTROMAGNETIC COMPLIANCE (EMC)

EMC testing at the vehicle level is required for EVs if they are to be legally sold and driven on public roads. Various regulatory bodies set standards and limits for electromagnetic emissions. EVs must comply with these standards to ensure that their electronic systems do not cause interference with other electronic devices and systems. The automotive industry and individual automobile manufacturers must meet a variety of those EMC requirements set. The requirements cover both conducted EMI and radiated EMI.

The International Special Committee on Radio Interference (CISPR) is one of the most adopted requirements. It is part of International Electrotechnical Standards (IEC). CISPR specifies that conducted or radiated emissions from a device must not exceed a certain level at specific frequencies. The emissions are quantified in dBuV. 'dBuV' stands for decibels relative to one microvolt. It is commonly used in the context of measuring and specifying levels of electromagnetic emissions in electronic systems. This measurement is used to describe the amplitude or strength of the interference signal at a specific frequency.

CISPR 25 is the requirement targeted for automotive applications. Compliance with CISPR 25 is crucial for automotive and other vehicle manufacturers to ensure that their products meet international EMC standards. It helps to prevent EMI issues that could affect the safety and performance of vehicles and ensures that they can be used without causing disruptions to other electronic systems. CISPR 25 is applicable to automotive components and modules. It uses one or two 5 μH/50 Ω artificial networks (also known as line impedance stabilization networks (LISN)) depending on the grounding configuration (Hegarty, 2018). The conducted emission measured from 150 kHz to 108 MHz should not exceed the limit listed in Figure 8.3. PK is for peak, Avg is for average. More detailed setup is available at Eaker (2022).

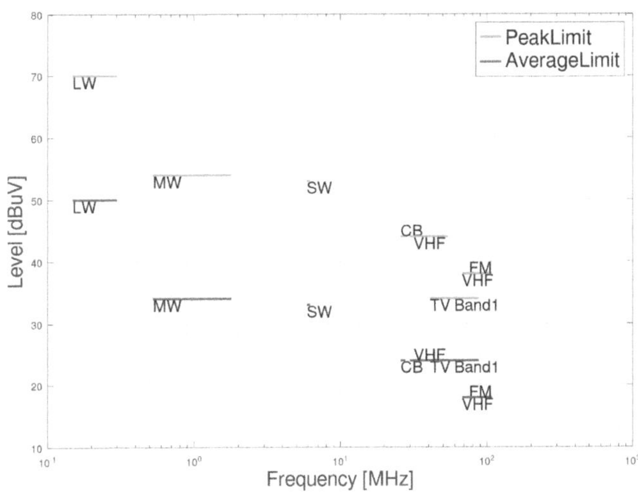

FIGURE 8.3 CISPR 25 Class 5 conducted emission limits.

The limits and methods of measurements specified in CISPR 25 are intended to protect onboard receivers from disturbances produced by components, such as a switching regulator in a power-supply design. CISPR 25 is more difficult than other typical requirements for industry applications, such as CISPR 22. The limits specified in CISPR 25 are very stringent, especially in the frequencies ranging from 68 to 108 MHz.

FACTORS AFFECTING EMI

EV electric drive systems are typically operated in switch mode with switching frequency ranges from tens of kHz to hundreds of kHz. As a result, large di/dt and dv/dt are produced by the power devices. This is one major root cause of EMI occurring in EVs. The large di/dt and dv/dt couple with parasitic inductance and capacitance via certain circuit loops, introducing EMI.

DEVICES AND GATE DRIVE CIRCUITS

MOSFETs or IGBTs are the key components for switching operations used in inverters and converters. It is the rapid switching of these components that introduces a high rate of change in current (*di/dt*) and voltage (*dv/dt*). The high rate of switching typically generates a broad spectrum of frequency components, including many HF harmonics, which are the source of EMI.

In switching mode (either converter or inverter) power systems, PWM switching of power devices (Si IGBT, Si MOSFET, or SiC MOSFET) creates discontinuous current that affects EMI in <30 MHz ranges. Fast rising and falling rates (dv/dt) of the switching devices impact EMI in 30–100 MHz. Additional ringing along switching edges due to parasitic inductances in the commutation loop affects EMI >100 MHz (Ramadass, 2023b; ROHM Semiconductors, 2018a). All thes fast switching creates HF harmonics that contribute to EMI.

The faster the rising and falling rates (switching speed), the higher the switching frequency, and the more likely it is to produce electromagnetic radiation in a higher frequency band. This can lead to more severe EMI. Therefore, a slower switching rate is desired to minimize EMI. On the other hand, fast switching is preferred to reduce switching loss. It is a trade-off. One key point (Zhang, 2020) for EMI is that 'the switching speed is of more importance than the switching frequency.' High di/dt and dv/dt could lead to more severe HF harmonics that extend beyond the fundamental frequency of the input or output signals. These harmonics cause electromagnetic emissions. HF harmonics are common since the inverter, OBC, and DC-DC switching waveforms are rarely sinusoidal. They are square waves, pulses, and inductive spikes with frequencies ranging from several to tens of kHz, including many HF harmonics. In addition, the wavelength of the electromagnetic waves also decreases with higher frequency, which potentially increases the possibility of making unintentional antennas out of circuit traces to induce more severe radiated emissions.

How fast a power device can switch is first determined by the type of device. Typically, a MOSFET has a much faster switching speed than an IGBT and can be used for higher switching frequencies. The reason is that MOSFETs are controlled by gate voltage only and can be switched on/off quickly, while IGBT control involves current, so it is slower. Secondly, gate drive circuits have a significant impact on the switching ramp rate. An optimized gate drive can quickly turn on/off the devices. For both Si IGBT and Si MOSFET, the gate drive needs to be optimized to achieve fast switching and minimal ripples that could introduce additional EMI. Switching speeds also vary with semiconductor material. Wide bandgap devices such as SiC and GaN enable power devices to operate at higher switching frequencies than traditional Si-based devices. As Husain (2024) pointed out, dv/dt is around 2–6 kV/μs for Si IGBT and 20–60 kV/μs for SiC MOSFET, which is about a tenfold difference. So dv/dt is extremely high for SiC compared to Si IGBT. This is one reason why EMI is more challenging for an SiC-based electric drive system.

Parasitic Components

In addition to devices, parasitic components associated with circuit layouts (loops) affect both dv/dt and di/dt. EMI-related dv/dt generally appears when the circuit couples to another circuit through parasitic capacitance. The amount of current produced will be proportional to the product of parasitic capacitance and dv/dt. Slow switching is sometimes preferred to reduce EMI but will not be preferred for thermal management since it leads to higher switching loss.

Some of the noise (dv/dt) produced by the switching node can go through parasitic capacitance, then go back onto the ground plane on the board, and become differential conducted noise. Some of the noise also passes parasitic capacitance to the ground in the environment or the heat sink ground, and is then radiated as CM noise. While charging and discharging the parasitic capacitance between high dv/dt nodes (normally power devices) and the ground, CM EMI noise is generated.

Rapid changes in current, characterized by high di/dt, induce voltage change across the inductive component and lead to EMI. EMI related di/dt is typically related to a current loop in a circuit. The current loop has certain parasitic inductance (*L*). Therefore, when a current passes through the loop, a magnetic flux will be created and further induce voltage in nearby loops (such as PCB traces). Both the magnetic flux and the voltage it introduces are proportional to Ldi/dt. The higher the parasitic inductances, the more severe the EMI will be generated. Therefore, the parasitic inductances need to be as small as possible to minimize radiated flux and EMI in OBC, DC-DC, and inverters.

Critical Parasitic Loop Inductances in Converters

Here, the critical loop in DC-DC is discussed. It not only covers HV to LV DC-DC used in EVs, but also it is an important portion of OBC. For buck converters (Hegarty, 2016), EMI noise is related to the parasitic inductances of the commutation loop and loop around gate drive circuits. The commutation loop is between the capacitor and the switches. The loop inductance increases switching-voltage ringing, affecting broad-band EMI in the 50–200-MHz range.

Critical Parasitic Loop Inductance in Inverter

Minimizing parasitic commutation loop inductance between power devices and DC capacitors is critical to ensure traction inverter performance. Figure 8.4 shows the commutation loop for a full bridge power module used in the traction inverter.

Parasitic inductance determines the voltage spike through the power devices during switching. Voltage overshoot typically occurs when the device turns off. It is a function of parasitic loop inductance L_p and current changing rate *di/dt*:

$$\Delta V_{overshoot} = L_p \times di/dt \tag{8.3}$$

Through parasitic inductance, high *di/dt* leads to high *dv/dt*. Larger commutation loop inductance causes higher voltage overshoot and higher *dv/dt*, more severe EMI, and higher power loss.

Layout optimization is needed to achieve minimal loop inductance and to minimize overshoot voltage during turn-off to ensure the safe operation of the devices. The commutation inductance loop includes the parasitic inductance of the DC capacitor, the parasitic inductance of the power module, and the connections between the DC capacitor and the power module. There are two aspects to minimizing loop inductance (Liu, 2020). First aspect is to design a low inductance power module, which is even more critical for SiC devices with high switching speed. State-of-the-art inverter power module inductance can be as low as 10 nH. ROHM developed TRCDRIVE pack™ series with two-in-one SiC molded modules for EV traction inverter applications. The module achieved a low inductance of 5.7 nH with two-layer busbar structure for critical wiring (ROHM Co. Ltd, 2024). The second aspect is to lower the inductance of the connection between the DC cap and power module. A laminated busbar connection is one typical solution.

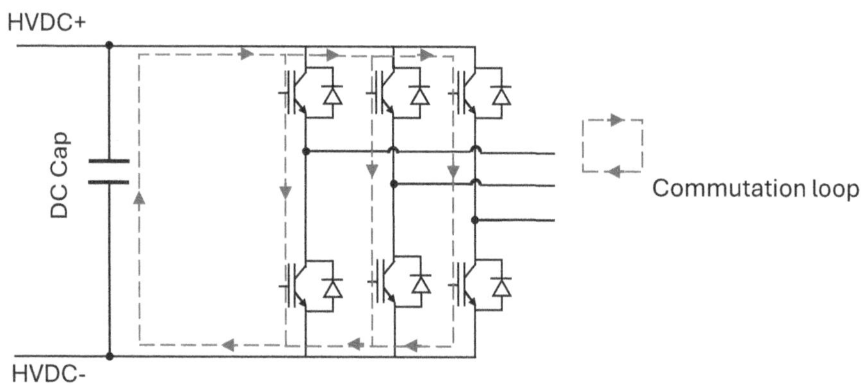

FIGURE 8.4 Commutation loop for full bridge inverter power module.

EMI IN POWER CONVERTERS

OBCs are a major EV component with EMI issues, normally related to power switching devices. In an OBC, the input current of the PFC is the main source of DM noise, while CM noise may come from both the PFC and DC-DC (primary and secondary). One major parasitic capacitance contributing to the CM noises is between the power devices and the heat sink which they attach to. The power devices are either from PFC or from primary and secondary DC-DC converters (seeFigure 5.2). The other is the parasitic capacitance between the primary and secondary windings of the transformer in the DC-DC converter portion (McDonald, 2022) . *Dv/dt* couples into the ground through parasitic capacitance. Noise current related to EMI injected via parasitic capacitance is normally CM noise. It is proportional to the product of parasitic capacitance and *dv/dt*.

For OBCs and DC-DC, discrete power devices such as TO247 are a common choice. The devices are attached to a heat sink via electrically insulated TIM. The heat sink is typically grounded (see Figure 8.5) so parasitic capacitance is formed between the devices and the heat sink. During switching, dv/dt from the devices will generate CM EMI noise current via capacitive coupling to the heat sink (Infineon, 2019):

$$i_{cm} = C_{par} \frac{dv}{dt} \tag{8.4}$$

Parasitic capacitance is

$$C_{par} \approx K\epsilon A/h \tag{8.5}$$

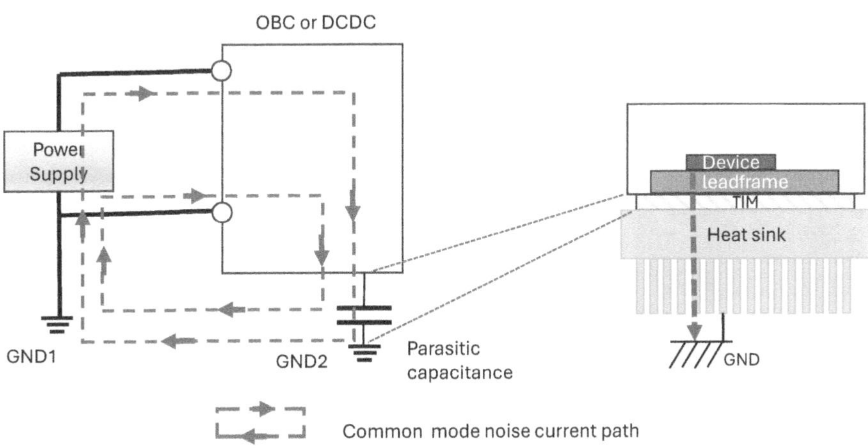

FIGURE 8.5 CM EMI current path for OBC DC-DC power devices to heat sink. The ground path may or may not go through the power supply, and the ground return path will be different.

The capacitance is related to package lead frame area (*A*), TIM bondline thickness (*h*), and relative permittivity of TIM (ϵ). Reducing the power device lead frame area and increasing bondline thickness will help to reduce the CM EMI current. However, this will make thermal performance worse (see Figure 8.4).

As Wyatt (2012) pointed out, EMI caused by CM current noise tends to be more of an issue than DM current noise induced EMI. One reason is, as previously discussed, CM is typically much higher than DM. Another reason is the CM current path is complex. It may come from different potentials from different ground planes. It may also come from unbalanced geometries.

EMI in Inverters and Motors

Another important source of conducted emissions is from traction inverters. This is due to the high-speed switching of power switches in the inverter. The high-speed switching introduces large *dv/dt* and *di/dt*, which in turn excites conducted emissions through parasitic capacitance or inductance of the circuits.

Figure 8.6 shows a CM EMI path for the powertrain system. Parasitic capacitance (C_{par_INV}) from the inverter, mostly from the power devices, to the chassis ground (GND) will form the path for the CM current to flow directly from the inverter to the ground. One important factor contributing to the CM EMI current is the parasitic capacitance from the power module to the heat sink, which is typically the ground. As discussed earlier, an electrically insulated layer (for example, the ceramic layer from DBC substrates) is sandwiched between power semiconductors and the heat sink, which creates the parasitic capacitance. Through this parasitic capacitance, CM noise generated by power devices will couple to the heat sink, then to the other components (Deboi, 2024). Another source is the intrinsic parasitic capacitance of the power devices (Zoeller, 2016). For IGBTs, parasitic capacitances include

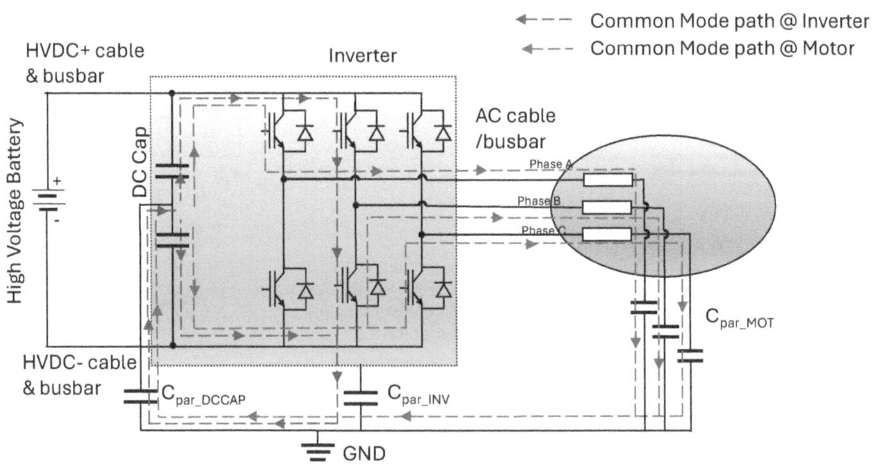

FIGURE 8.6 CM noise current path for EV powertrain system (Wang, 2022).

gate to emitter capacitance, gate to collector capacitance, and collector to emitter capacitance.

CM noise current I_{CM} (Zhang, 2021) can be determined by parasitic capacitance (C_{par}) and voltage change rate ($\frac{dV}{dt}$)

$$I_{CM} = C_{par} \times \frac{dV}{dt} \tag{8.6}$$

Parasitic capacitance is the function of three capacitances.

$$C_{par} = f(C_{device}, C_D, C_L) \tag{8.7}$$

C_{device}: device parasitic capacitance. For SiC MOSFET, it is drain to source capacitance

C_D: freewheeling diode capacitance

C_L: parasitic capacitance due to layout. For example, the capacitance between the device and heat sink

The CM paths at the motor side are more complex. One path is through parasitic capacitance between the stator winding and the stator housing ground. The other is from the stator winding to the rotor via parasitic capacitance, then to the ground via stator-rotor capacitance or through the motor bearing (Herdin, 2021). The parasitic capacitance between the traction motor winding and stator housing ground can be significant and is typically between 5 and 25 nF (I. Husain, 2021). If the inverter is away from the motor, the AC connections using shielded cables introduce additional parasitic capacitance. Integrating the inverter to the motor, as many EVs have done, has the benefit of reducing EMI as a result of minimized parasitic capacitance from AC connections to the ground. Another benefit is that without AC cables, the emission from the cables is minimized (Jenkins, 2020).

In EV traction motors, the CM noise current due to high dv/dt may introduce bearing current that can reduce reliability (Morya, 2019; Herdin, 2021). Bearing in motors normally has very thin layers of lubricant that sit between two parts, forming a high parasitic capacitance. The capacitance may have random breakdowns from time to time due to the difficulty in maintaining the thin layer of lubricant and unperfect bearing. This random breakdown can generate high random peak CM current noise (Zhang, 2021). There are two possible paths for bearing current, one from the bearing to the stator frame and then to the ground, and the other circulating between the rotor, shaft, bearing, and stator frame (Adabi, 2007).

The DM EMI path of the inverter is the parasitic commutation loop between DC link capacitors and the switches (see Figure 8.7). DM voltage in the motor appears across two different AC phases of the motor. It introduces leakage current between motor windings from different phases. This will not only introduce DM EMI noise, but also introduce winding insulation degradation/damage over time and make DM EMI worse.

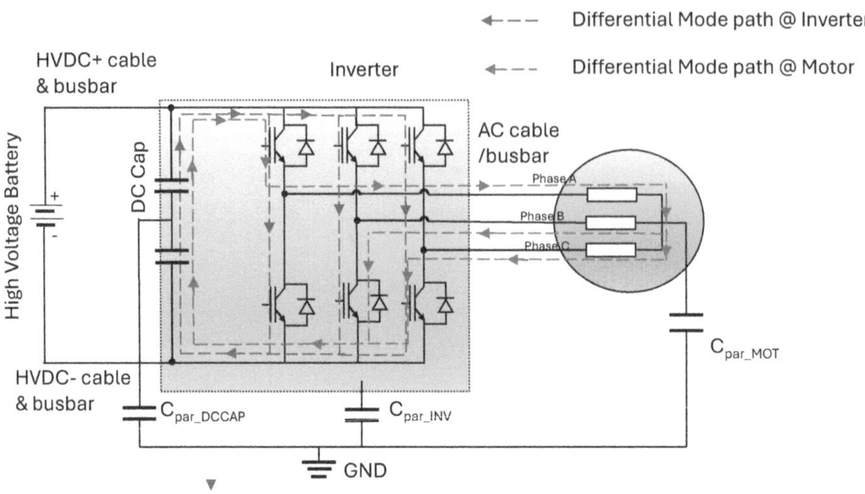

FIGURE 8.7 DM noise current path for EV powertrain system (Wang, 2022).

Resonance

Resonance (harmonics) is a critical factor that resulted in an unacceptable level of EMI. More than 80% of EMC problems are related to resonance (Quarktwin, 2023). EMI resonance occurs when the EMI noise frequency matches the natural resonant frequency of some part of the electric drive system.

In power electronics systems, signals have different frequencies, including PWM switching harmonics as previously discussed. HF noise signals will attenuate by inductors in series and capacitors in parallel. However, resonance will occur when the frequency of the signal matches the inductor-capacitor (LC) circuit. The LC circuit could be related to the cables, PCB traces, or parasitic inductance and parasitic capacitance of power electronic circuits. The LC circuit can be excited around its resonance frequency, resulting in unacceptable levels of EMI noise. One example (TI, 2017) occurred to the buck converter when parasitic energy excited a resistor-inductor-capacitor (RLC) resonance. It resulted in amplified ringing and voltage overshoot. In this example, the inductors are the parasitic inductance related to the commutation loop, and the capacitors are the parasitic capacitance of the switching devices.

Cable

In addition to devices and circuit layout (loops and parasitic components), cable is also another important factor affecting EMI. The cable affects EMI in several ways (Emisoftware blog, 2023):

- It serves as the path for conducted EMI
- It serves as an antenna to radiate EMI
- It picks up EMI noises when exposed to electromagnetic fields

The interaction between cables and EMI is complex and depends on various factors such as cable construction, length, termination, and the nature of the signals being transmitted.

As an example, fast switching of power devices in an inverter introduces conducted emissions. The conducted emissions can be transmitted either through DC input cables to the battery or AC output cables to the motor. The conducted EMI can couple with the vehicle chassis metal chassis, which leads to CM current and radiated EMI.

Cable Radiation (Antenna)

Cables can act as antennas, especially when they are long and unshielded. The antenna effect appears when the cable length is comparable to the wavelength of the interfering signals. Table 8.1 lists the frequency of the signal and its corresponding wavelength. An antenna is excited most effectively when the length of the antenna is ½ or ¼ of the wavelength of the source frequency. Cables longer than 0.375 m pretty much can be treated as antennas for 200 MHz signal if not shielded.

Cable Resonance

Like circuit resonance, cable resonance introduces a higher voltage and current and leads to a much higher electromagnetic field at resonant frequency.

In actual applications, most cables can resonate at certain frequencies (Emisoftware blog, 2023) since almost all the cables are not perfectly terminated at both ends. When a cable's electrical parameters (inductance, capacitance, and resistance) combine to generate a resonance, the cable will be resonant to signals at that specific frequency. This can result in increased voltage levels and current flow at the resonant frequency.

Other EMI Sources in EVs

LV signals in EVs also contribute to EMI due to their HF. LV signal frequency range for LV signals used in EV powertrain modules is discussed in Zhang M. (2021): Frequency range for LV switch mode power supply is 150 KHz–2 MHz, 10 KHz–2.5

TABLE 8.1
Wavelength at Different Switching Frequencies

Frequency	Wavelength	½ Wavelength	¼ Wavelength
30 MHz	9.99 m	4.99 m	2.50 m
50 MHz	6.00 m	3.00 m	1.50 m
100 MHz	3.00 m	1.50 m	0.75 m
200 MHz	1.50 m	0.75 m	0.375 m

MHz for controller area network (CAN) communication, and 200–600 MHz for microcontrollers. Those LV components, especially control circuits, are placed in close proximity to or even inside HV power electronic components. Therefore, LV signals may be interfered with by HV signals if not carefully designed.

EMI MITIGATION

To predict and mitigate EMI requires a thorough understanding of the system, including the components, PCB layout, and signal paths, to identify potential sources of electromagnetic emissions and susceptible areas (Texas Instruments, 2017). EMI mitigation is typically addressed in several aspects: component selection, circuit design, grounding, filtering, and shielding. Grounding and shielding are two important EMI mitigation techniques. They are often addressed at different levels: circuit level, PCB board level, and system level (Sushmitha, 2022).

To identify and minimize critical loops and relevant component parasitic capacitance/inductance is one aspect of minimizing EMI. This is related to PCB layout and system packaging. Another aspect is to identify critical EMI sources and related frequency ranges/relevant harmonics and minimize them if possible. This can be realized via filtering or circuit optimization. These important aspects related to EV electric drive systems are discussed in more detail next.

Grounding

A significant percentage of EMI issues that occurred in the field are related to 'grounding.' The ground is the reference point for voltage. Grounding in EVs is very important to control HF noise linked to EMI. Multi-point grounding is better for HF EMI control because it minimizes ground impedance and reduces the voltage potential for unwanted voltage differences across a system at high frequencies. There is chassis ground, analog ground, and digital ground. It is very important to separate 'more noisy' return current from 'less noisy' return current by the proper separation of the ground for those currents. How to design ground properly has been studied extensively (Peterson, 2018; Zhang, 2021).

EMI Filters

EMI filters are the most common approach to mitigate EMI conducted noise. They are designed for specified frequencies or bands of frequencies. An EMI filter uses passive components such as capacitors and inductors.

Based on the types of EMI, there are CM filters and DM filters (see Figure 8.8). CM EMI is typically the priority for EMI mitigation. It is present in both signal and return paths, in the same phase with respect to each other. CM filtering typically involves a CM inductor in series with both the signal line and the return line and/or capacitors to the ground. Mutual inductance will maximize the impedance to CM noise. As a result, the CM noise current will be attenuated. DM filtering involves placing capacitors between the signal and return lines and/or an inductor in series

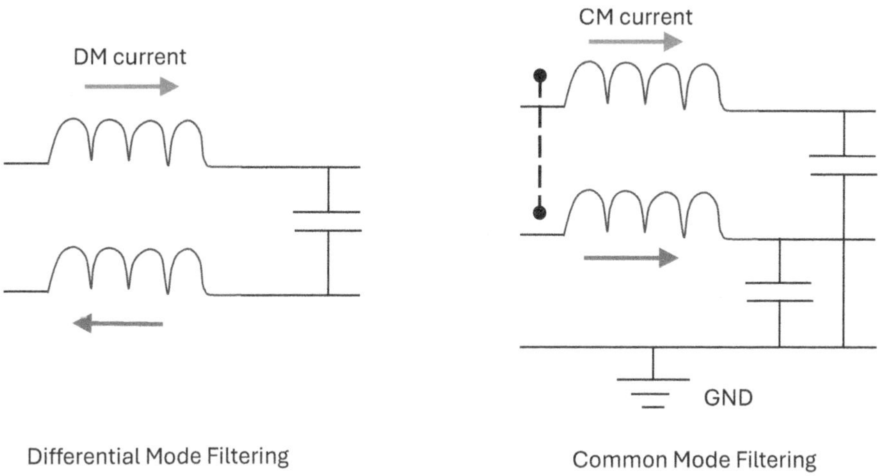

FIGURE 8.8 Schematics of DM and CM filtering (Interference Technology, 2023).

with either the high or low side of the line (Interference Technology, 2023) (see Figure 8.8).

In traction inverters, the DC link capacitor helps to reduce current ripple and voltage fluctuation and helps to minimize EMI. In addition, CM noise is also minimized by the CM choke and Y-capacitors. X capacitors are used to suppress DM noise. The EMI components are part of the inverter on the DC input side. The DC current from the battery will pass through the EMI filter and DC capacitor before passing through the switches. This breaks the CM noise current path (Texas Instructments, 2018). Similar considerations also applied in the EMI filter for the OBC and DC-DC converter.

Hesener (2010) and Coilcraft (2011) explained how to design EMI filters:

The input data needed is:

f_{line}: line frequency
V_{min}: minimal RMS voltage
I_{max}: maximum RMS load current
f_{swmin}: lowest switching frequency

The filter can be determined by following the steps.

Step 1. Determine the design impedance: $Z = V_{min}/I_{max}$
Step 2. Determine filter topology
Step 3. Determine the required attenuation at the lowest problem frequency f_{sw_min}. This is determined by simulation or measurement
Step 4. Decide filter cutoff frequency f_{cut}. This is typically based on the attenuation needed at the concerned frequency, with a margin of a few

dB, while also ensuring minimal attenuation at the line frequency. Normally, $f_{cut} \geq 10 f_{line}$

Step 5. Determine L and C of the filter using the following equations

$$L = Z / 2\pi f_{cut} \quad C = 1 / 2\pi f_{cut} Z$$

PCB and System Layout

PCB layout and component placement are critical for EMI mitigation (Laird, 2021). For OBC and DC-DC converters, reducing parasitic capacitance near switching devices is important. This is sometimes addressed by reducing trace dimensions on the PCB without scarifying thermal performance too much (Texas Instructments, 2018). Another good practice is to minimize critical parasitic commutation loop inductance in PCB layout design. The loops are normally related to the input capacitor and switching devices. In a traction inverter, the commutation loop area between the DC link capacitor and switching devices needs to be minimized.

Good practice includes placing a ground plane next to the power plane on the PCB to act as a shield. The ground plane provides the least impedance path for the return current. This helps to minimize EMI (Texas Instruments, 2021). Using full ground plane or ground grids (Sushmitha, 2022) is also good practice. Another option is to minimize cross talk and coupling between adjacent traces. It is recommended to keep signal and power ground connections separate except for differential traces. HV portions should be completely isolated from the LV portion if possible (Taranovich, 2023). Faraday's cage can be used to isolate the EMI noisy portion from the EMI clean portion.

EMI Mitigation in Cable Harness

Cables and harnesses carrying analog and digital signals can pose serious EMI issues due to their parasitic capacitance and inductance. EMI mitigation methods include shielding source components and corresponding wiring harnesses. Conducted and radiated EMI can be minimized in longer harnesses by adding a filter. Careful planning of the wiring harness by arranging lower power circuitry closer to the signal source and higher power interfering circuitry closer to the load will help (Taranovich, 2023). Improved grounding techniques, shielding harnesses, and connecting the shield to the vehicle body ground will help to reduce EMI. HV cables connecting the traction inverter and motors are typically shielded (Zhang, 2021).

Resonance

Identifying and minimizing resonance is important for EMI mitigation. This can be performed in a systematic way (Herrick, 2009):

- Identify the frequency generated by the sources under different operating conditions

- Identify the natural resonant frequency of the transfer path or resonant structure
- If source frequency matches the resonance frequency of the transfer path
- Adjust source frequency
- Adjust resonance frequency of the transfer path
- Add filtering

An antenna can also be treated as a structure resonance. Based on Table, ¼ wavelength of a 200 MHz signal is around 0.375 m. In an electric drive system, it is preferred to have the cable or HV harness short. As discussed in the earlier chapter, the trend of integrating several HV systems together will help to mitigate EMI issues via shorter HV cables/connections. The examples are most EVs, where traction inverters are integrated with the motor. AC cable connections between the inverter and the motor is replaced by short busbar connections. This reduces power line length and EMI at the same time.

Once the critical frequency, the source, or the path contributing to the critical frequency is identified, it is easier to determine the critical factors that contribute to radiation and mitigation strategy.

EMI MEASUREMENT METHODS AND TOOLS

One typical type of EMI testing is compliance testing. The goal is to ensure EMI generated by the electronic system under specified test conditions does not exceed criteria specified by EMC standards such as CISPR 25. In addition, EMI testing is also used for EMI troubleshooting to identify specific components or areas that contribute to EMI.

The HF current probe is a tool used to measure HF current flowing in the conductors. It is used in conducted EMI testing to measure HF current flowing on conductors connected to the devices. It can also be used in radiated EMI testing to determine emission sources and to assess EMI fields of interest (Chen, 2010). HF current probes are typically used for near-field measurement to determine hot spots without physically disturbing the circuit.

A far-field EMI probe is used to measure emissions at the far field. The probe includes various antennas depending on the specific frequency and directional characteristics needed. It can be used for compliance testing as well as to determine critical frequencies of interest for troubleshooting.

LISNs are an important testing component in EMI testing and measurement, mainly for conducted emissions measurements. They are typically specified in various EMC/EMI test standards such as CISPR. It is a low pass filter placed between the power source and unit under test (UUT). LISNs provide a precise impedance (typically 50 ohm), allowing repeatable EMI measurements. They block unwanted HF noise from the power source while letting the power frequency pass to the unit. They provide a radio frequency noise measurement port for conducted emissions.

EMC testing on the vehicle level for EVs is required. The testing can be very costly and time-consuming. Using EMI simulation is a promising but challenging option to minimize EMC testing effort.

BIBLIOGRAPHY

Adabi, J. e. (2007). Leakage Current and Common Mode Voltage Issues in Modern AC Drive Systems. *Power Engineering Conference, 2007.* IEEE.

Chen, C. e. (2010). A Study of PCB EMI Measurement and Simulation. *Asia-Pacific International Symposium on Electromagnetic Compatibility.* Beijing, China: IEEE.

Coilcraft. (2011, 08 07). *Common Mode Filter Design Guid.* From https://www.coilcraft.com/getmedia/9c231e30-04a2-4463-b679-c38b99b2669e/doc191_CMFiltDesign.pdf

Deboi, B. (2024, Jul 03). *Power Modules: One Shortcut to EMI Compliance.* From Wolfspeed: https://www.wolfspeed.com/knowledge-center/article/power-modules-one-shortcut-to-emi-compliance/

Eaker, M. e. (2022, July). *Understanding CISPR25 Current Probe and Voltage Method for Conducted Emissions.* From https://www.ti.com/lit/ab/slla569a/slla569a.pdf?ts=1732496743479

Emisoftware blog. (2023). *Cabling Affects Electromagnetic Interference in Complex Ways.* From https://www.emisoftware.com/blogs/cabling-affects-electromagnetic-interference-complex-ways/

Hegarty, T. (2016). *Reduce Buck-Converter EMI and Voltage Stress by Minimizing Inductive Parasitics.* From Texas Instruments Analog Application Journal: https://www.ti.com/lit/an/slyt682/slyt682.pdf?ts=1732650584067&ref_url=https%253A%252F%252Fwww.google.com%252F

Hegarty, T. (2018, Feb.). *An Overview of Conducted EMI Specifications for Power Supplies.* From Texas Instruments: https://www.ti.com/lit/wp/slyy136/slyy136.pdf?ts=1732502535794&ref_url=https%253A%252F%252Fwww.google.com%252F

Herdin. (2021, Nov 09). From Troubleshooting Common-Mode-EMI in Electric Drive Installation: https://eepower.com/technical-articles/troubleshooting-common-mode-emi-in-electric-drive-installations/#

Herrick, C. (2009, Sep). *Fundamentals of EMI.* From IEEE: https://ewh.ieee.org/r6/scv/emc/archive/092009Herrick.pdf

Hesener, A. (2010). *Electromagnetic Interference (EMI) in Power Supplies.* From Fairchild Semiconductor Power Seminar 2010–2011: https://emcfastpass.com/wp-content/uploads/2017/04/Electromagnetic-Interference-EMI-in-Power-Supplies.pdf

Husain, I. (2024). *WBG Power Electronics Driving High-Speed Electric Machines for Electric Vehicles.* From https://poweramericainstitute.org/wp-content/uploads/2021/11/8.-High-Speed-Drives_2021-Husain.pdf

Husain, I. (2021). Electric Drive Technology Trends, Challenges, and Opportunities for Future Electric Vehicles. *IEEE* (pp. 1039–1059). IEEE.

Infineon. (2019, Jul 29). *Infineon Application Note: TO-247-3 Advanced Isolation: Infineon's New Fully Isolated Discrete Package.* From https://www.infineon.com/dgdl/Infineon-TO-247-3_Advanced_Isolation-ApplicationNotes-v01_01-EN.pdf?fileId=5546d4625fe367840160021aeb971416

Interference Technology. (2023, Feb 16). *EMI Filtering 101: Understanding the Basics.* From https://interferencetechnology.com/emi-filtering-101-understanding-the-basics/

Jenkins, J. (2020, Jan 7). *EMC for EVs: Understanding Electromagnetic Compatibility.* From https://chargedevs.com/features/emc-for-evs-understanding-electromagnetic-compatibility/

Laird. (2021, Apr 16). *A Basic Overview of EMI Shielding Techniques & Concepts.* From https://www.mouser.com/pdfDocs/abasicoverviewofemishieldingtechniquesconcepts_mouser.pdf?srsltid=AfmBOorY_ZalpCBL30aEpA_1h5INTfUpo7sn0Vru3zsBYJu8IDywzhAj

Liu, G. E. (2020, Feb 01). *Recent Advances and trend of HEV/EV-Oriented Power Semiconductors – An Overview.* From IET Power ElectronicsVolume 13, Issue 3 p. 394–404. https://ietresearch.onlinelibrary.wiley.com/doi/full/10.1049/iet-pel.2019.0401

McDonald, B. (2022). *Addressing EMI Conducted Emissions Challenges in Electric Vehicles with GaN-Based OBCs.* From https://www.ti.com/lit/an/slyt831/slyt831.pdf?ts=1732566708703&ref_url=https%253A%252F%252Fwww.google.com%252F

Moon, M. E. (2015). Modeling and Analysis of Return Paths of Common Mode EMI Noise Currents from Motor Drive System in Hybrid Electric Vehicle. *2015 Asia-Pacific Symposium on Electromagnetic Compatibility (APEMC).* Taipei, Taiwan: IEEE.

Morya, A. K. (2019). Wide Bandgap Devices in AC Electric Drives: Opportunities and Challenges. *EEE Transactions on Transportation Electrification*, *5*(1), 3–20.

Peterson, Z. (2018, Sep 25). *All About Ground Reference and Chassis Ground in Electronics Design.* From https://resources.altium.com/p/stay-grounded-digital-analog-and-earth-ground-pcb-layout

Quarktwin. (2023, 03 24). *How to Solve the EMC Problem Caused by Resonance.* From https://www.quarktwin.com/blogs/temperature%20sensitive%20component/how-to-solve-the-emc-problem-caused-by-resonance/46

Ramadass, Y. (2023a). From Introduction to EMI: Standards, Causes and Mitigation Techniques: https://www.ti.com/document-viewer/lit/html/SSZT173

Ramadass, Y. (2023b, Dec). *Time-Saving and Cost-Effective Innovations for EMI Reduction in Power Supplies.* From https://www.ti.com/lit/wp/slyy200a/slyy200a.pdf?ts=1732647447573&ref_url=https%253A%252F%252Fwww.electronicdesign.com%252F

ROHM. (2024). *ROHM Semiconductors.* From Silicon-carbide (SiC) Power Devices: https://www.rohm.com/products/sic-power-devices

ROHM Co. Ltd. (2024, Jun 11). *ROHM's New TRCDRIVE Pack with 2-in-1 SiC Molded Module: Significantly Reduces the Size of xEV Inverters.* From ROHM Semiconductor: https://www.rohm.com/news-detail?news-title=2024-06-11_news_trcdrive-pack&defaultGroupId=false

ROHM semiconductors. (2018a, Mar 8). *Differential (Normal) Mode Noise and Common Mode Noise–Causes and Measures.* From Rohm Tech Web: https://techweb.rohm.com/product/nowisee/6899/

ROHM Semiconductors. (2018b, May 24). *Noise Occurring in Switching Power Supplies.* From https://techweb.rohm.com/product/nowisee/7121/

Sushmitha. (2022, Dec 6). *https://www.protoexpress.com/blog/7-pcb-design-tips-solve-emi-emc-issues/.* From https://www.protoexpress.com/blog/7-pcb-design-tips-solve-emi-emc-issues/

Taranovich, S. (2023, Jan 20). *How to Mitigate EMI in Electric and ICE Vehicles.* From Electronicdesign.com: https://www.electronicdesign.com/technologies/power/white-paper/21258511/electronic-design-how-to-mitigate-emi-in-electric-and-ice-vehicles

Texas Instructments. (2018, Apr 16). *Automotive EMI Reduction Techniques, Applications, and Solutions.* From https://www.ti.com/video/5771546750001

Texas Instruments. (2017, May 10). *Understanding EMI and Mitigating Noise in DC/DC Converters.* From https://www.ti.com/video/5429498088001

Texas Instruments. (2021, Apr 16). *Identifying the Key Features of an Automotive DC/DC Converter.* From https://www.ti.com/video/5083871639001

TI. (2017, May). *Understanding EMI and Mitigating Noise in DC/DC Converters.* From https://www.ti.com/video/5429498088001

Wang, K. E. (2022). Modeling of System-Level Conducted EMI of the High-Voltage Electric Drive System in Electric Vehicles. *IEEE Transactions on Electromagnetic Compatibility, 64*(3), 741–749.

Wyatt, K. (2012). *EMC Directory & Guide: The HF Current Probe: Theory and Application.* From https://interferencetechnology.com/wp-content/uploads/2012/04/Wyatt_NA_DDG12.pdf

Zhang, M. (2020, Jul 16). *Demystifying EMC in an Electric Vehicle's Drive Unit.* From https://interferencetechnology.com/demystifying-emc-in-an-electric-vehicles-drive-unit/

Zhang, M. (2021, Jan). *EMC Design Techniques for Electric Vehicle Powertrain Modules.* From IN Compliance: https://incompliancemag.com/emc-design-techniques-for-electric-vehicle-powertrain-modules/

Zoeller, C. E. (2016). Influence of Parasitic Capacitances of IGBT Inverter on Insulation Condition Monitoring of Traction Machines Based on Current Signal Transients Analysis. *18th European Conference on Power Electronics and Applications.* Karlsruhe, Germany: IEEE.

9 Trends in EV Electric Drive Systems

ABSTRACT

One significant trend for EV electric drive systems is the integration of the inverter, motor, and transmission into a single electric drive system. This significantly reduces size, weight, and cost while enabling better performance. It also simplifies the assembly and makes modularization easy. Electric drive systems are inherently modular. One basic electric drive system unit can be shared across the product portfolio with minimal design changes, which minimizes the time-consuming and costly process of designing separate electric drive systems for each new vehicle model. Doubling the power capability can be achieved by using two motors and doubling the battery capacity. The approach is widely adopted in the EV industry to reduce costs and accelerate innovations.

Wide adoption of EVs is enabled by technological advancements in electric drive systems. These technologies enable better performance, higher power density, and lower costs. Several key technological advancements have been discussed in previous chapters: better thermal performance like double side cooling and new package technologies including Ag sintering to increase reliability. This chapter covers both the adoption of wide bandgap devices and 800 V HV, which enable more efficient and higher power density electric drive systems that better meet customers' needs.

INTEGRATION AND MODULARIZATION

One significant trend in EV electric drive systems is the integration of the electric motor, inverter, and transmission into a single compact electric drive unit. One example is the Lucid electric drive unit (shown in Figure 9.1). This integration reduces weight, size, and cost while simplifying vehicle assembly and potentially improving efficiency. The unit achieved best-in-class power density in terms of electric drive units: >18 kW/L and >7 kW/kg.

Another trend is the modular design of electric drive units. It allows scalability across different vehicle classes. Volkswagen's MEB platform, General Motors' Ultium platform, and Hyundai Motor'sE-GMP are examples of scalable electric drive systems that can be shared among various vehicle models. In the MEB platform (Volkswagen Canada, 2022), real wheel drive is the standard option. All-wheel drive is based on rear-wheel drive with an additional electric motor in the front axle. Battery housing is easy to reconfigure to accommodate different number of battery modules. This allows the same base drive system to be shared across a wide variety

DOI: 10.1201/9781003569961-9

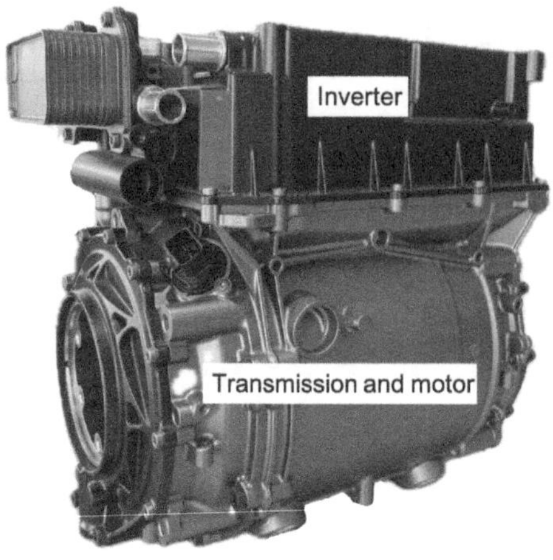

FIGURE 9.1 Lucid electric drive unit assembly, including motor, inverter, and transmission (Photo by Author).

of vehicle types, including compact cars, SUVs, and vans. The MEB platform includes at least eight production models such as the Audi Q4 e-tron, Volkswagen ID.4, and Ford Explorer EV. It consolidates electronic controls and reduces the number of microprocessors. Another example is Tesla. The Tesla model Y shares a significant percentage of its components with the Model 3, where the powertrain is a major portion. As discussed earlier, development, especially testing, is a major cost factor for EVs. Sharing the same electric drive system across the entire EV model portfolio significantly reduces development costs.

Electric drive systems are inherently modular. The power and torque of an EV can be adjusted by simply modifying the number and size of the motor and battery pack. One example is the Tesla Plaid rear drive unit (Kane, 2021). Plaid cars can provide a combined power output of 1020 horsepower. Instead of completely redesigning its drive unit to achieve high power, it is equipped with three drive units, which are based on the Model 3 drive unit: one in the front and two in the rear. The rear drive units are symmetric two drive units attached back-to-back to each other (See Figure 9.2). Each drive unit is very similar to one Model 3 rear drive unit, which includes one motor and one inverter. In comparison, it is very difficult to modularize ICEs. An ICE for a small car is very different from the engine of a truck. It is impractical to use two ICEs from a small car as the engine for a truck.

SIC

SiC devices have gained significant market share in the EV industry. They are becoming a preferred option for power semiconductor devices for high power (>10 kW), HV (600–1700 V), HF (20–200 kHz) applications (ROHM Co. Ltd, 2024; Sanken Electric, 2024; Lovati, 2021).

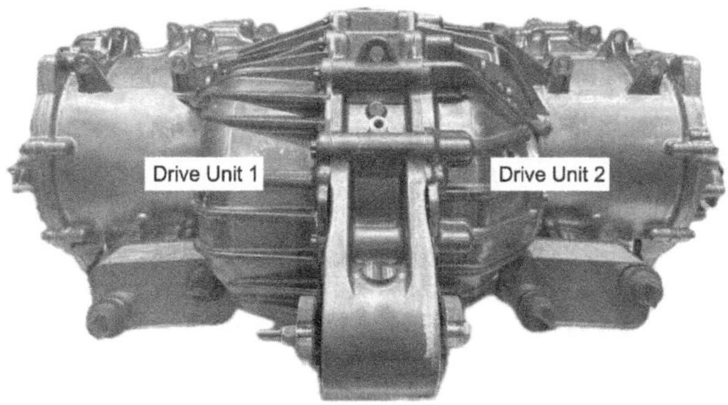

FIGURE 9.2 Tesla Plaid Rear Drive Unit, which has two drive units back-to-back, each unit includes one motor and one inverter (back of the image), and one shared transmission (Photo by Author).

Why SiC MOSFET?

SiC is an ideal choice for HV applications (600–1700 V). There are two characteristics that allow SiC to operate at higher electric fields and voltages without breaking down. Firstly, the bandgap of SiC is almost three times that of Si (see Table 9.1). Secondly, dielectric breakdown electric field of SiC is 10 times that of Si because silicon and carbon are tightly bonded in SiC, and its crystal structure is more stable than that of single crystal silicon. This enables much thinner SiC devices under the same breakdown voltage rating than Si.

SiC MOSFETs are better suited for high switching frequency applications than Si IGBT. Electron Saturation velocity of SiC is twice that of Si. Therefore, SiC can be switched on and off faster than Si and can achieve lower switching loss. SiC has around two orders of magnitude lower on resistance per unit area than Si. This leads to lower conduction losses and higher efficiency (Nisch, 2018).

SiC IGBT has significantly high switching loss at higher frequencies (>20 kHz). In contrast, SiC MOSFET has low switching loss even at high frequencies due to low

TABLE 9.1
Comparison of Material Properties of Si and SiC (ROHM Co. Ltd, 2024; Zhang, 2019)

	Si	4H-SiC
Bandgap (eV)	1.12	3.2
Dielectric Constant	11.7	9.66
Dielectric Breakdown Electric Field (mV/cm)	0.3	3
Electron Saturation Velocity (10^7 cm/s)	1	2
Bulk Electron Mobility (cm^2/Vs)	1350	720
Thermal Conductivity (w/mK)	1.5	4.5
Figure of Merit	1	500

turn-on resistance and fast operational capability. Therefore, most SiC switches are MOSFETs.

The wider bandgap of SiC (see Table 9.1) allows it to have a much lower thermal leakage current than Si at high temperatures (Callana, 2011). Wider bandgap means more energy is needed for electrons to move from the valence band to the conduction band, reducing intrinsic carriers at high temperatures. SiC can be operated at 200°C or even higher. In comparison, Si power devices are normally rated at 150–175°C maximum. Additionally, SiC has similar ability as Si to control p-type and n-type regions necessary for power device fabrication (ROHM, 2016).

The advantages of SiC over Si are mainly HV, lower loss, and high temperature operation properties. Studies by Toyota researchers (Hamada, 2015) have shown that SiC MOSFETs and diodes have lower conduction losses over a wide current range than Si IGBTs and diodes. At the same time, they still maintain faster switching performance. According to ROHM (2024), SiC MOSFET significantly improves traction inverter efficiency in the high torque and low rotational speed range over Si IGBT and results in a 6% reduction in electricity consumption. The advantages of SiC become more obvious when comparing Si IGBT devices and SiC MOSFET devices at the system level. The Infineon CoolSiC MOSFET (MOUSER ELECTRONICS, 2024) used in traction inverters increases battery utilization by 5–10%. It also increases system power density with size reduction of up to 80%. It has lower conduction loss in light load conditions compared to Si IGBT. Hain (2019) has compared efficiency of an 800 V Si IGBT-based inverter and an 800 V SiC MOSFET-based traction inverter. SiC MOSFET has much lower turn-off losses than Si IGBT. At light load (<~54% maximum output current), the SiC MOSFET-based inverter has lower conduction loss than the Si IGBT-based inverter. At high load, the SiC MOSFET-based inverter may have higher conduction loss. Under standard WLTC driving profiles, the Si IGBT inverter reaches an average efficiency of 94.4%. In comparison, the SiC MOSFET inverter can achieve 97.3% efficiency. Thus, using SiC MOSFETs can reduce the loss by 50%. Ultimately, the newest SiC MOSFETs can achieve average efficiency of 98.7% and improve the driving range by 7%. For the SiC MOSFET inverter power module, it can be combined with the external diode typically required in Si IGBT inverter power modules. A Schottky barrier diode (SBD) can be integrated into SiC MOSFETs, which reduces the active area of the die. This leads to a significant reduction in power module area and loss (Liu, 2020; Daryanani, 2025).

Another case is in OBC or DC-DC applications, as Hammerschmidt (2017) has shown, at high switching frequencies (>10 kHz), SiC MOSFET is more efficient. In the study, at 40 KHz, SiC can use 120% more load current than Si IGBT. For OBC and DC-DC applications, higher switching frequencies will help to reduce the size of passive components, including inductors and transformers. Comparing a DC-DC converter with a SiC MOSFET at 160 KHz to the same system with Si IGBT at 25 kHz, weight is reduced from 7 to 0.9 kg and volume is reduced by 85%.

In addition, the higher the efficiency of SiC-based power electronic systems also leads to less cooling needs. For example, if the efficiency of a 20 kW power system increases from 97% to 98%, it corresponds to a power loss from 0.6 to 0.4 kW. It is expected to significantly reduce the cooling system demand.

The main challenge of SiC over Si is the higher device cost. The cost of the SiC substrate is an important factor (Hamada, 2015). SiC is a compound of silicon and carbon. In order to grow SiC crystals, high-purity raw materials and complex synthesis processes are needed. Additionally, it is more difficult to produce SiC wafers since SiC is much harder. Lastly, the yield of SiC wafers is generally lower than that of Si wafter because the SiC crystal structure intrinsically has more defects than the Si wafer.

State of Art of SiC Packages

Tesla is the pioneer in using SiC in EV traction inverters. The inverter uses 24 pairs of SiC MOSFET rated at 650 V each (Barbarini, 2018). Its overall package design (see Figure 9.3) has several innovations beyond the use of SiC devices. A Cu lead frame instead of an Al wire bond is used, which increases current capacity. Silver sintered die attach has replaced the traditional solder attach. The discrete package uses a ceramic-based substrate as the baseplate instead of a copper lead frame, allowing the package to be directly sintered to the heat sink. This brings the thermal resistance per unit area of the package in line with the value of a multichip power module. These approaches, together with SiC enabled higher power density, higher temperature capability, and improved inverter reliability (Barbarini, 2018). Based on Munro & Associates' report (Munro & Associates, 2020), this leads to a higher power-to-weight ratio than other EV inverters (see Figure 9.4).

The SiC power module from the Mitsubishi Electric for EV/HEV applications have shown a reduction of power loss by 62% compared with the Si power module (Wada et al, 2017). At the same time, with the same packaging, the module size is decreased by ~50%. Rohm (Editorial Staff, 2019; Mashaly, 2018) found replacing Si IGBT at 16 kHz with SiC at 24 kHz in a Formula E inverter can increase power by

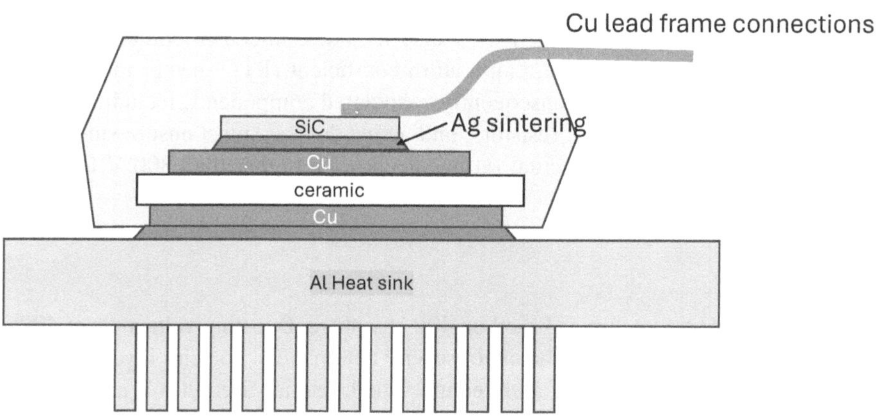

FIGURE 9.3 Cross-sectional view of Tesla Model 3 inverter power module stack-up.

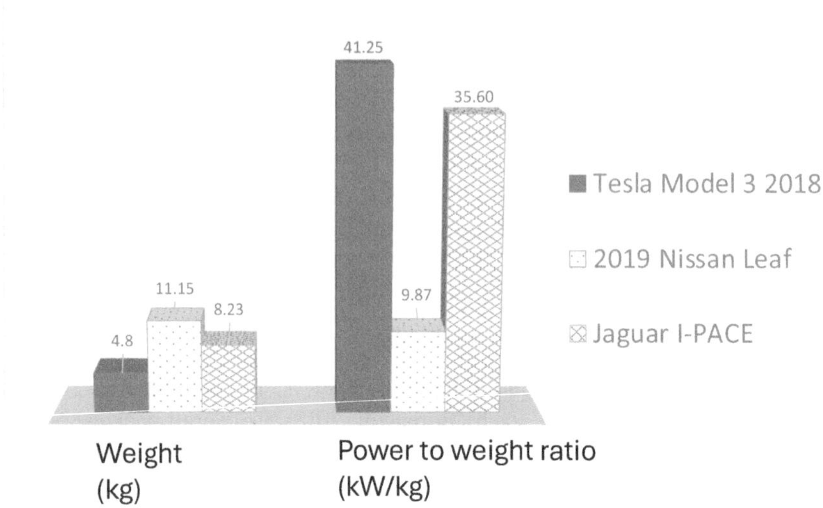

FIGURE 9.4 Weight and power density comparison between the 2018 Tesla Model 3 inverter, the 2019 Nissan Leaf inverter, and the 2019 Jaguar I-Pace inverter (Munro & Associates, 2020).

10% and increase efficiency ~1% at both maximum efficiency and maximum power conditions. At the same time, overall weight is reduced from 15 to 9 kg, and volume is reduced from 14 to 10 L. In addition, driving range increases from 159to 177 km.

Another driver for SiC inverters, the shift of HV electrical architecture from 400 to 800 V, is discussed next.

800 V HV ELECTRICAL ARCHITECTURE

In 800 V HV architecture (see Figure 9.5), HV DC bus voltage connecting various subsystems is rated at 800 V, which includes the battery, the motor, the inverter, the DC-DC converter, and the OBC. It also includes other HV components such as the compressor, the positive temperature coefficient (PTC) heater module, and the power distribution unit. Consequently, associated components, including cables, connectors, fuses, capacitors, resistors, and active devices, must ensure safe operation at 800 V voltages. The actual ratings need to be higher than 800 V to ensure reliability over their lifetime.

Introduction

Typical EV HV architecture is based on 400 V voltage. Boosting voltage from 400to 800 V or higher is a preferred choice for several automakers. Porsche Taycan was one of the first EVs to adopt 800 V architecture. The Lucid air used 900 V architecture. The shift to 800 V or higher voltage architectures is part of a broader industry effort to enhance the charging capabilities, performance, and efficiency of EVs.

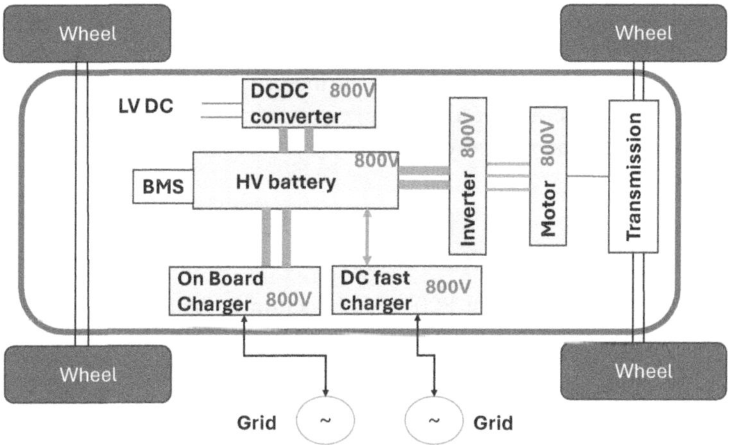

FIGURE 9.5 EV 800 V HV architecture.

One main motivation for 800 V is fast charging. A major hurdle to the wide adoption of current EVs is that under typical 400 V DC architecture, charging time is too long compared to gasoline vehicles. Increasing the output power by either increasing the voltage or charging current can mitigate this issue. Under typical 400 V charging, normal current is limited to roughly 400–500 A due to charging connectors overheating. If the charging power is 160 kW under 400 V and 400 A, it will take at least 1 hour to fully charge a 160 kW EV. This is obviously not acceptable for normal users. Boosting the voltage from 400 V to 800 V theoretically can double output power and reduce the charging time by half and better meet normal customers' expectations.

The alternative solution adopted by Tesla is to increase charging current. The charging power capability of the Tesla supercharger can reach 250 kW by boosting the charging current to 615 A. However, this poses a thermal challenge to charging cables due to the high current. Typical DC fast charging cables and connectors will be too hot to handle the heat loss generated by the high current without significantly increasing cable cross-sections (Jenkins, 2019). Tesla designed liquid-cooled cables to solve this challenge.

In comparison, assuming 800 V and 500 A charging, it will theoretically take only 7.5 minutes to reach 80% SOC from 30% SOC for a 100 kWh battery. Therefore, 800 V is a feasible solution adopted by more and more automotive OEMs, including Porsche, Hyundai, GM, and Delphi. For both the Porsche and Hyundai Ioniq EV, charging time from 5 to 80% is around 20 minutes. For Porsche, the maximum charging power is 225–270 kW. The charging time can be as quick as 22.5 minutes from 5 to 80% with a DC fast charger and 93 minutes with Level 2 50 kW charging (Porsche Fremont, 2020). The Hyundai Ioniq 5, based on an 800 V E-GMP, can be charged from 10 to 80% (over 200 miles) in 18 min (Crosse, 2021).

Another key benefit for 800 V is the significant reduction of current under the same power rating, enabling the use of higher power density and more efficient power electronics and motors. To transfer the same amount of power, doubling the voltage from 400 to 800 V means reducing current by half and reducing resistive losses (I^2R) fourfold, which is a major contributor to the power electronic components and motor losses. As a result, power electronic components, such as inverters and converters, can operate more efficiently with reduced power loss under 800 V architecture.

800 V SiC Inverter

The most significant impact of 800 V architecture is that SiC power devices are becoming the mainstream choice. For an 800 V inverter, the devices need to be rated at 1200 V at the power stage. The SiC device is the key enabler for 800 V powertrain architecture. Even though there are 1000 V rated Si IGBT devices, they are typically used at lower switching frequencies in the single-digit kHz range (Jenkins, 2021).

Hyundai uses Infineon's CoolSiC power module for their 800 V-based E-GMP. The higher efficiency of SiC has enabled a vehicle range increase of more than 5% (Lombardo, 2021). It is reported that a 2% power-stage efficiency gain with the adoption of SiC corresponds to an additional 7 miles of range per battery (Slovick TI, 2023). Replacing a 400 V Si IGBT inverter with an 800 V SiC inverter could result in a range increase of 5% for a light-duty EV (Allca-Pekarovic, 2023). The BorgWarner 800 V SiC inverter has reduced power losses by 40–70% as compared to silicon-based inverters, depending on the drive cycle (Spendlove, 2021). Lucid Air's 900 V inverter uses SiC power modules from Wolfspeed. Hitachi provided an 800 V inverter based on its double-sided cooling power module in the Porsche Taycan EV. Compared to its previous generation inverters, Hitachi's 800 V solution delivers 2.7 times higher power density (Hitachi, 2019).

800 V Battery

To increase battery pack voltage from 400 to 800 V is straightforward: doubling the cells in series. Use the Tesla S battery case study in Chapter 4 as an example. The configuration is 96 cells in series and 74 in parallel. The nominal cell voltage is 3.8 V, and the cell current is ~4.89 A. The battery system voltage is ~365 V. Now the battery pack needs to provide voltage of ~700 V with the same capacity.

To double the pack voltage, the number of cells in series should be doubled:

$$96 \times 2 = 192$$

Therefore, battery pack voltage is

$$V = 3.8V \times 192 = 729.6V$$

To keep the same capacity, the total number of cells should be the same, so the number of cells in parallel is reduced by half: 74/2=37

Now, the battery pack current is reduced.

$$I = 4.89A \times 37 = 181A$$

To change the battery pack voltage from ~400 to ~800 V while maintaining the same capacity, we just need to double the cells in series and reduce the cells in parallel to half.

800 V Motors

Running at a higher pack voltage (800 V) means the speed and power of the motor can be higher without increasing current and torque. Higher voltage results in higher motor speed (up to 20,000 RPM) and higher power under the same current and torque (Jenkins, 2021). Since motor size is proportional to the torque specification, higher voltage motors can accommodate higher power without significantly increasing the motor size and weight (Jolley, 2021). In reality, increase in AC loss in the motor will occur due to higher speed, but this is not expected to be significant.

One impact of the 800 V motor is the significant reduction in copper winding cross-section area and higher motor power density. Theoretically, as the voltage increases from 400 to 800 V, it can handle almost twice the power under the same current. Alternatively, it can handle the same power with significantly less current, leading to reduced copper winding loss from joule heating (I^2R). Ultimately, this means higher efficiency and a significant reduction in motor dimensions and weight. 800 V or higher motor voltage rating will result in a smaller motor size than a 400 V motor with the same power rating. One example is the traction motor used in the Lucid air (see Figure 9.1). With a higher power rating (900 V), it is much smaller than typical EV motors. According to Halvorson (2022), the Lucid traction motor was able to achieve a power density of 21.3 HP/kg (668.82 HP), well below 10.0 HP/kg (470 HP/47 kg) for a German EV and 8.4 HP/kg (336.84 HP/40.1 kg) for an American EV. Similarly, BorgWarner's (BorgWarner, 2021) latest 800 V electric motor, the High Voltage Hairpin 320 motor, is capable of 97% peak efficiency and also enables a higher power density and a reduced charging time.

800 V OBC

With voltage increased from 400 to 800 V, the OBC can achieve a higher power rating with the same current or significantly better efficiency with the current drop by half. An 800 V OBC also requires HV (800 V) rated power switches. BorgWarner used SiC devices for its bi-directional 800 V OBC (The EV Report, 2023). Its power rating is approximately 20 kW.

800 V Capacitors

For 800 V inverters and OBCs, 'The first big change is that electrolytic capacitors might not be practical to use because 450 V is the highest working voltage that is commonly available.' Film and ceramic dielectric capacitors are proven to be better choices above 400 V or so (Jenkins, 2021). Obviously, the DC link capacitor in the inverter needs to be changed from a 400 to 800 V rating. This likely will result in a different capacitance value, efficiency, and cooling requirement.

Cables

One of the biggest advantages of 800 V EV architecture is the thinner and lighter cables due to lower current. Jenkins (2021) has given an example of 350 kW DC fast charging; a 12 kg/27# 5 m cable would be sufficient with 800 V. In comparison, a 37 kg/81# 5 m cable is needed for 400 V charging. The reduction in cable weight from 37 to 12 kg is significant. 800 V does require thicker insulation, but the weight increase from insulation is expected to be minor.

BIBLIOGRAPHY

Allca-Pekarovic. (2023). Loss Modeling and Testing of 800 V DC Bus IGBT and SiC Traction Inverter Modules. *IEEE Trans. on Transportation Electrification*, *10*(2).

Barbarini, E. (2018, Jun). *STMicroelectronics SiC Module: Tesla Model 3 Inverter*. Retrieved from www.systemplus.fr: https://web.archive.org/web/20210325095228/https://www.systemplus.fr/wp-content/uploads/2018/06/SP18413-STM_SiC_Module_Tesla_Model_3_Inverter_sample-3.pdf

BorgWarner. (2021, Feb 10). *BorgWarner Launches 800-volt Electric Motor for the Commercial Vehicle Segment*. Retrieved from https://www.borgwarner.com/newsroom/press-releases/2021/02/10/borgwarner-launches-800-volt-electric-motor-for-the-commercial-vehicle-segment

Callana. (2011, May 11). *Technology Behind the SiC MOSFET and Applications*. Retrieved from Power Systems Design: https://www.powersystemsdesign.com/articles/cost-competitive-sic/37/4309

Crosse, J. (2021, Feb 23). *Why 800V Charging Is the Next Step for Mass EV Adoption*. Retrieved from https://www.autocar.co.uk/opinion/new-cars/why-800v-charging-next-step-mass-ev-adoption

Daryanani, S. (2025, Mar 19). *Advances in Integrating Schottky Barrier Diodes with SiC MOSFETs*. Retrieved from https://www.powerelectronicsnews.com/advances-in-integrating-schottky-barrier-diodes-with-sic-mosfets/

Editorial Staff. (2019, May 17). *Benefits of SiC MOSFET in Powertrain Inverter for Automotive Applications*. Retrieved from https://www.powerelectronicsnews.com/benefits-of-sic-mosfet-in-powertrain-inverter-for-automotive-applications/

Hain, S. E. (2019). Evaluation of 800V Traction Inverter with SiC-MOSFET versus Si-IGBT Power Semiconductor Technology. *PCIM Europe 2019*. Nuremberg, Germany: VDE.

Halvorson, B. (2022, Sep 13). *How Lucid Leaps Past Tesla with Smaller Motors*. Retrieved from https://www.greencarreports.com/news/1137141_lucid-motors-more-power-dense-easier-to-build-than-tesla

Hamada, K. E. (2015). SiC—Emerging Power Device Technology for Next-Generation Electrically Powered Environmentally Friendly Vehicles. *IEEE Transactions on Electron Devices*, *62*(2), 278–285.

Hammerschmidt, C. (2017, Jun 14). *The Potential of Silicon Carbide (SiC) for Automotive Applications*. Retrieved from EEnewsEurope: https://www.eenewseurope.com/en/the-potential-of-silicon-carbide-sic-for-automotive-applications/

Hitachi. (2019, Oct 18). *Hitachi Automotive Systems Starts Mass Production of High Voltage and High Output EV Inverter, the First in the World*. Retrieved from https://www.hitachi.com/New/cnews/month/2019/10/191018.html

Jenkins, J. (2019, Nov 19). *Why did Porsche Go to the Trouble of Designing an 800 V Taycan EV?* Retrieved from Charged EV: https://chargedevs.com/newswire/why-did-porsche-go-to-the-trouble-of-designing-an-800-v-taycan-ev/

Jenkins, J. (2021, Feb 24). *High-Voltage EV Battery Packs: Benefits and Challenges. More Voltage, More Better?* Retrieved from https://chargedevs.com/features/high-voltage-ev-battery-packs-benefits-and-challenges-more-voltage-more-better/

Jolley, D. (2021, Feb 15). *Shifting to 800-Volt Systems: Why Boosting Motor Power Could Be the Key to Better Electric Cars.* Retrieved from https://www.youris.com/mobility/traffic_management/shifting-to-800-volt-systems-why-boosting-motor-power-could-be-the-key-to-better-electric-cars.kl

Kane, M. (2021, Oct 14). *See The Tesla Plaid's Dual-Motor Rear Drive Unit.* Retrieved from https://insideevs.com/news/540692/tesla-plaid-drive-rear-unit/

Liu, G. E. (2020). Recent Advances and Trend of HEV/EV-Oriented Power Semiconductors – An Overview. *IET Power Electronics*, *13*(3), 394–404.

Lombardo, T. (2021, May 11). *Infineon Introduces SiC Six-Pack Power Module for EV Traction Inverters.* Retrieved from https://chargedevs.com/newswire/infineon-introduces-sic-six-pack-power-module-for-ev-traction-inverters/

Lovati, S. (2021, Mar 17). *10 Things to Know About Silicon Carbide (SiC).* Retrieved from https://www.powerelectronicsnews.com/10-things-to-know-about-sic/

Slovick TI. (2023, Jun 15). *Maximize EV Range with Higher-Efficiency Traction Inverters.* Retrieved from Electronic Design: https://www.electronicdesign.com/tools/learning-resources/whitepaper/21267152/texas-instruments-maximize-ev-range-with-higher-efficiency-traction-inverters

Mashaly, A. (2018, July 16). *Utilizing SiC Traction Technology Inverters in Automotive Applications.* Retrieved from EEPower: https://eepower.com/technical-articles/utilizing-sic-traction-technology-inverters-in-automotive-applications/#

Mouser Electronics. (2024). Retrieved from Infineon Technologies CoolSiC™ MOSFETs: https://www.mouser.com/new/infineon/infineon-coolsic-mosfets/?srsltid=AfmBOopW_sE0DTM-gWzP-AmeufiMYUYJmAy6L-FfW0q3WS9hUwIozO6M

Munro & Associates. (2020, Dec 10). *Munro & Associates Report Details the Ins and Outs of EV Inverters.* Retrieved from https://vasa.org.au/munro-associates-report-details-the-ins-and-outs-of-ev-inverters/

Nisch, A. E. (2018). *Effects of a SiC TMOSFET Tractions Inverters on the Electric Vehicle Drivetrain.* Retrieved from https://www.infineon.com/dgdl/Infineon-Effects_of_a_SiC_TMOSFET_tractions_inverters_on_the_electric_vehicle_drivetrain-Editorials-v01_00-EN.pdf?fileId=5546d4627600a6bc017601596e5162b8

Porsche Fremont. (2020, Dec 20). *What's the Porsche Taycan Charge Time?* Retrieved from https://www.porschefremont.com/porsche-taycan-charge-time/

ROHM. (2016, Sep 8). *What Is Silicon Carbide?* Retrieved from Rohm Tech Web: https://techweb.rohm.com/product/power-device/sic/3669/

ROHM. (2024). *ROHM Semiconductors.* Retrieved from Silicon-carbide (SiC) Power Devices: https://www.rohm.com/products/sic-power-devices/sic-mosfet#supportInfo

ROHM Co. Ltd. (2024, Jun 11). *ROHM's New TRCDRIVE pack with 2-in-1 SiC Molded Module: Significantly Reduces the Size of xEV Inverters.* Retrieved from ROHM Semiconductor: https://www.rohm.com/documents/11303/13026034/Introduction_NewRelease_TRCDRIVEpack_e.pdf/99157c86-82de-a19f-94ba-c7ae2e7af186?t=1718085675617

Sanken Electric. (2024). *Compound Semiconductors: What Are "SiC" and "GaN"?* Retrieved from Sanken Electric Co., Ltd.: https://www.semicon.sanken-ele.co.jp/en/guide/GaNSiC.html

Spendlove, T. (2021, Nov 29). *BorgWarner Will Supply 800 V SiC Inverter to German automaker.* Retrieved from https://chargedevs.com/newswire/borgwarner-will-supply-800-v-sic-inverter-to-german-automaker/

The EV Report. (2023, Nov 3). *BorgWarner Announces 800V Charger for North American OEM*. Retrieved from https://theevreport.com/borgwarner-announces-800v-charger-for-north-american-oem

Volkswagen Canada. (2022, Apr 8). Retrieved from Architecture and functions of Volkswagen's MEB chassis for the ID. family: https://www.youtube.com/watch?v=Fpnn8VShd0U

Wada, F., et al. (2017). 6-in-1 Silicon Carbide (SiC) MOSFET Power Module for EV/HEV Inverters. *PCIM Asia 2017.* Shanghai: IEEE.

Zhang, Y. (2019). Comparison Between Competing Requirements of GaN and SiC Family of Power Switching Devices. *IET Power Electronics*, *13*(3), 394–404. Retrieved from https://iopscience.iop.org/article/10.1088/1757-899X/738/1/012004/pdf

Index

800 V, 144
800 V SiC inverter, 146

A

Acceleration, 15
Accelerometer, 113
AC frequency, 3, 30; *see also* fundamental frequency
Aerodynamic loss, 10
Ag sintering, 98–100; *see also* silver sintering
Air gap, 37
Al wedge wire bonding, 86–88
Ancillary loss, 10
Angular speed, 2; *see also* towing

B

ase speed, 11
Battery, 1; *see also* EV battery
  adjusted capacity, 8; *see also* required capacity
  battery capacity, 1
  Lithium ion battery, 1
Battery loss, 47
Battery Management System, 50–51; *see also* BMS
Battery pack, 51
Battery resistance, 47
  alternating current internal resistance, 47; *see also* ACIR
  direct current internal resistance, 47; *see also* DCIR
Battery system isolation fault, 50
Battery system packaging, 51
Bearing, 23
Bearing current, 129

On board Charger, 1, 4, 64–65; *see also* OBC
  Bidirectional OBC, 4
Bottom cooling, battery, 55

C

Cable, 130, 134
  cable radiation, 131; *see also* also antenna
  cable resonance, 131
Cell module, battery, 51
Charging, 17, 46
  level, 1 charging, 17
  level, 2 charging, 17
  level, 3 charging, 17
CISPR, 25, 123
CM noise current, 121, 129
Coefficients of thermal expansion, 85, 92; *see also* CTE
Coffin Manson based model, 96
Common mode EMI, 62, 121; *see also* CM
Component vibration resistance, 108
Conducted EMI, 120
Controller, 63
Coolant, 38
  automatic transmission fluid, 38–39; *see also* ATF
  housing jacket cooling, 39
  shaft cooling, 40
  spray cooling, 39–40
  through lamination cooling, 41
  water/ethylene glycol, 38–39; *see also* WEG
Creep, 95
Creep energy density, 97
Creep strain, 97
CTE mismatch, 96
Current sensor, 62
Cu wire bonds, 98
Cycles to failure, 94, 97, 113

D

Damping, 116
DC capacitor, 62
DC-DC converter, 1, 4, 65–66
Desired range, 8
di/dt, 124–125
Die attach, 85–86, 88, 91
Die attach failure, 93
Differential mode EMI, 62, 121; *see also* DM
Dipole antenna, 122
Direct Bonded Copper, 74; *see also* DBC
Direct current, 1; *see also* DC
Discharging, 46
Discrete power electric packages, 69; *see also* also discrete packages
  DPak/D2Pak Discrete Power Electronic Packages, 71
  TO220, 69
  TO247, 69
DM noise current, 121
Double side cooling power module, 79–80
Drivetrain loss, 10

Driving range, 8–10
Dual active bridge converter, 64; *see also* DAB
dv/dt, 124–125

E

E_in, 4
E_out, 4
Eddy current, 26
Efficiency, 4, 17–18, 31
efficiency map, 32–34
efficiency of motor, 31–33
Electric drive systems, 1, 139
Electric field radiation strength, 122
Electromagnetic Compliance, 123; *see also* EMC
Electromotive force, 10; *see also* EMF
back EMF, 11
EMI, 119
EMI filters, 62
EMI filters for OBC, 65
EMI mitigation, 132
CM filtering, 133
DM filtering, 133
EMI filters, 62, 64, 65, 132–134
grounding, 132
EV Battery
anode, 45
battery capacity, 48
battery cell, 45, 47
battery voltage, 48
cathode, 45
C-rate, 49
cylindrical battery cell, 46–47
electrodes, 45
electrolyte, 46–47
maximum, 30-sec discharge current, 49
maximum continuous current, 49
maximum continuous power, 50
peak power, 50
pouch battery cell, 46–48
prismatic battery cell, 46–48
state of charge, 49, 50; *see also* SOC
Excitations, vibration, 106; *see also* source; vibration

F

Far field EMI probe, 135
Fast charging, 145
Fastener, 115
Flux density, 10, 28
FR4, 73
Frequency separation, vibration, 114
Fundamental frequency of PCB, 108, 113

G

Gate drive board, 62–63
Gear ratio, 3
Gradeability, 13
Ground plane, 128

H

Heating, battery, 55
Heat sink attach, 85–86, 88, 91
Heat sink attach failure, 92
Heat spreading, 77
Heat transfer coefficient, 6; *see also* HTC
HF current probe, 135
High frequency harmonics, 124
High voltage, 1, 4; *see also* HV
Horsepower, 13; *see also* HP

I

IEC, 60068-2-6 sinusoidal vibration testing, 116
IGBT, 62, 67
Inductor-Inductor-Capacitorˇconverter, 64; *see also* LLC
Insulated Metal Substrate, 73; *see also* IMS
Integration, electric drive system, 135
Interconnects, 85
Interconnect vibration fatigue, 108
Interior permanent magnet, 25; *see also* IPM
Intermetallics, 90
Internal combustion engine, 1; *see also* ICE
Inverter, 3, 61–64; *see also* traction inverter
Inverter cooling design, 63
ISO, 16750-3, 110–111
ISO, 19453-3, 110–111

J

Junction temperature, 70, 75

L

Laminations, 26
LC circuit, 130
Lead free solder, 88–89
Leakage current, 129
Lifetime analysis, 94
Liquid cooling, motor, 37
LISNs, 135
Loop antenna, 122
Loss, power semiconductor devices, 66
conduction loss, 66
switching loss, 66

Loss, SiC MOSFET, 142; *see also* also SiC MOSFET efficiency
Lower Voltage, 1, 4; *see also* LV
Low frequency harmonics, 117

M

Magnetic field, 22
Maximum continuous current, 12
Maximum motor speed, 30
Mission profile, vibration, 111–113
Modal analysis, 117
Modular design, electric drive system, 135–136
Moment of inertia, 2
MOSFET, 62, 67
Motor, 1–3, 22
- AC induction motors, 26–27
- AC motors, 10
- motor speed, 3
- permanent magnetic synchronous motors, 2, 22
- torque, 27–28
- traction motor, 2

Motor loss, 34
- core loss, 35
- eddy current loss, 35
- electrical loss, 34
- friction loss, 34
- hysteresis loss, 35
- mechanical loss, 34
- winding loss, 34

Motor shaft, 23
Multichip power module, 69, 74–81

O

On-chip interconnects, 85–86
Overcharging, 53
Overdischarging, 53

P

Parasitic capacitance, 127–129
Parasitic components, 125
Parasitic loop inductance, 125–126
Peak current, 12
Performance characteristic, EV, 8
Permanent magnetics, 23
Pole number, 3
Pole number, 3, 22
Power, 1, 13–14, 29
- AC power, 1
- DC power, 1
- maximum continuous power, 13, 28, 29, 63
- mechanical power, 1
- peak power, 11, 13, 28, 29, 63; *see also* also maximum power

Power capability of power electronic components, 68
Power cycling test, 93–95
Power devices, 61–62
Power electronic components, 61
Power electronic package, 69–79, 85, 93
Power loss, 4
Power spectral density, 109; *see also* PSD
Pressfit, 100
Printed Circuit Board, 73; *see also* PCB
Pulse width modulation, 3; *see also* PWM

R

Radiated electromagnetic field, 122
Radiated EMI, 121
Random vibration, 107
Range of vehicle, 1; *see also* vehicle range
Reflow soldering, 89–90
Resonance, EMI, 130
Resonance, vibration, 107
Resonant frequency, 114
Ribbon bonding, 88
Road excitation, 110
Rotor magnetics, 25

S

Saliency ratio, 25
Saturation, 29
SiC, 136–144
SiC MOSFET, 137
SiC packages, 143
Side cooling, battery, 55
Sinusoidal excitation, 109–111
Slippage, 3
S-N fatigue model, 113
Solder joint fatigue, 96–98
Solder void, 90
Speed, 30
Stator, 22
Stiffness, 114–115
Surface permanent magnet, 25; *see also* SPM
Switching frequency, 3

T

Thermal cycling test, 93–95
Thermal design of power electronic components, 68, 79; *see also* also thermal design of power semiconductor devices
Thermal interface material, 70; *see also* TIM
Thermal management, battery, 54–57
Thermal management, motor, 34

Thermal management system, 4
Thermal resistance, 5
  thermal resistance for conduction, 5
  thermal resistance for convection, 5
Thermal resistance, motor, 36
Thermal resistance from junction to ambient, 70
Thermal resistance from junction to case, 70
Thermal runaway, battery, 52–54
Thermal time constant, 29
Thermo-mechanical fatigue, solder joint, 91–94
Thermo-mechanical fatigue model, 96
Tire diameter, 3
Top speed, 16
Torque, 2, 10, 28; *see also* motor torque
  maximum continuous torque, 10, 29
  peak torque, 10, 12, 13, 27, 29; *see also* maximum torque
Towing capacity, 3
Transfer molded package, 101
Transient thermal response of power electronic packages, 78
Transmission, 3
  multispeed transmission, 4
  single speed transmission, 3–4

V

Vehicle speed, 3
Vibration analysis, 113; *see also* also nCode
Vibration proof capacitor, 116
Vibration proof component, 113
Vibration reliability, 106
Vibration resistance, 116; *see also* also vibration resistant
Vibration stress amplitude, 113

W

Wide bandgap, 142
Wide open throttle, 15; *see also* WOT
Winding, 24
  concentrated winding, 25
  distributed winding, 25
  hairpin winding, 24
  stranded winding, 24
Wire bond failure, 93
Wire bonds, 86–88; *see also* wire bonding

X

X-Cap, 62

Y

Y-Cap, 62